EASTERN TIMES

CONTENTS

Introduction	3
J69 on Lowestoft shed	4
Stratford to Chingford via Hall Farm Junction	5–20
In defence of Sir Thomas Bouch	21–27
Neville on the South Bank	28–39
The magnificent Claud Hamiltons	40–51
Hatfield Logs, 30th March 1961	52–61
The Thompson Class L1 2-6-4 Tank Locomotives	62–73
51E Stockton Shed	74–79
The Headshunt	80

EASTERN TIMES • ISSUE 8

May 1945 • Three well-dressed gentlemen are present at the launch of Thompson's Class L1 No. 9000 which took place at Liverpool Street. I wonder if the tallest gentleman on the left is Edward Thompson? See pages 62–73 for more on the Class L1 2-6-4 tank locomotives.
Photo: © Transport Treasury

© Images and design: The Transport Treasury 2025. Design and Text: Peter Sikes

ISBN: 978-1-917776-23-2

First published in 2025 by Transport Treasury Publishing Ltd., 16 Highworth Close, High Wycombe HP13 7PJ.

The copyright holders hereby give notice that all rights to this work are reserved. Aside from brief passages for the purpose of review, no part of this work may be reproduced, copied by electronic or other means, or otherwise stored in any information storage and retrieval system without written permission from the Publisher. This includes the illustrations herein which shall remain the copyright of the copyright holder.

Copies of many of the images in EASTERN TIMES are available for purchase/download. In addition the Transport Treasury Archive contains tens of thousands of other UK, Irish and some European railway photographs.

www.ttpublishing.co.uk or for editorial issues and contributions email: tteasterntimes@gmail.com

Printed in the UK by Short Run Press, Bittern Road, Sowton Industrial Estate, Exeter EX2 7LW.

INTRODUCTION

A striking photo of a Holden J69 starts us on another journey through various areas of the Eastern Region in issue 8.

Our first article is from Dave Brennand who takes us on a trip from Stratford to Chingford, taking in Hall Farm Junction. There are some wonderful period photographs, the majority of them illustrating how tidily the infrastructure was maintained, accompanied by clear uncluttered trackbeds, something of a rarity today.

Ian Lamb forms a defence of Sir Thomas Bouch, who is unfortunately remembered for the disaster that befell the Tay Bridge. This masked his achievements including designing what in effect was the first 'roll-on/roll-off' ferry and his bridge building on the Stainmore line, these lasting until physically dismantled during the closure of many routes of the railway system in the 1960s.

Our next article features a friendly rivalry from the north and south banks of the Humber. We feature many photographs from the Neville Stead Collection in the pages of Eastern Times. In this two-part feature Paul King selects images from the county of Lincolnshire and accompanies them with in-depth captions.

David Cullen takes a look at the construction of the Great Eastern Railway Holden-designed Class S46 4-4-0 which took place in 1900. The first loco was numbered after the year of construction and was named *Claud Hamilton*. The initial locomotive was followed by a further 40 locomotives built between 1900 and 1903 in four batches of ten. The first three images in the article are from the LNER era but no date or location details were recorded by the photographer, if any reader can recognise where the images were takenthen please drop me a line by email.

In issue 7 Geoff Courtney completed his record of trainspotting at Ilford. Not to be content with spotting at his home station Geoff also took his notebook to other Eastern Region locations. In this issue he recalls a trip to Hatfield on the East Coast Main Line.

The photo to the left is a taster to our next offering, a feature on the Thompson Class L1 2-6-4 tank locomotives. These are a favourite of regular contributor Dave Brennand, and as usual the text is accompanied by some great images.

Finally, and appropriately in the year of Railway 200, we feature a location that was prominent at the beginning of the modern railway – Stockton. For this issue we are not delving into the history of the area but are just taking a quick look at the locomotive shed and its environs.

Thank you to the readers that took the time to write in, their comments are included 'Headshunt' and are much appreciated. So, if you have anything you would like to comment on, or an article you think would be an interesting addition to Eastern Times, please don't hesitate to send it to me at *tteasterntimes@gmail.com*

Eastern Times is published three times a year, and is available on a subscription service. To sign up use the contact details below and ensure your copy is automatically sent to you every four months. *https://ttpublishing.co.uk/transport-books/*, email *admin@ttpublishing.co.uk*, or call us on **01494 708939**.

PETER SIKES, EDITOR, EASTERN TIMES
email: tteasterntimes@gmail.com

Front cover: A glorious 1st November 1958 at Stratford in the heart of London's East End, where we witness Doncaster-built N7/3 No. 69727 0-6-2T basking in the sunshine, whilst BR Standard 4 2-6-0 No. 76031 is in the shadows. Both were allocated to 30A in the late 1950s. The N7 would only see another two years' service, mainly on NE London suburban trains, whereas the BR Standard saw further use on the Southern Region after steam's demise at Stratford in September 1962. The N7 has been freshly painted. Its brasswork and side rods have been polished. It is not known whether this was for a special working, or just Stratford's everyday pride in the appearance of its locomotives. *Photo: D. Brennand Collection*

J69 ON LOWESTOFT SHED

22nd May 1957
Holden Class J69/1 0-6-0T No. 68565 pictured moving off-shed at Lowestoft (32C).
The J69/1 was a 1902 development of the J67 class with larger water tanks and firebox.
Photo: R. C. Riley © Transport Treasury

STRATFORD TO CHINGFORD VIA HALL FARM JUNCTION

TEXT AND PHOTOS (UNLESS STATED): DAVE BRENNAND

The very first trains to travel over the new Chingford branch in North East London would not have traversed the Hackney Marshes from Hackney Downs via Clapton as many would suspect. Instead, trains originated from Lea Bridge Road and headed to the first station on the branch at St. James Street, which opened in 1870, along with Hoe Street and terminated at Shern Hall Street. The latter was sited in a cutting between the current day Walthamstow Central (Hoe Street) and Wood Street.

It would be a further two years before the more direct route from Liverpool Street via Bethnal Green and Hackney Downs was completed. Lea Bridge Road station on the early Stratford to Broxbourne route was opened in 1840 by the Northern & Eastern Railway; it was renamed Lea Bridge in 1841. The temporary terminus at Shern Hall Street only lasted for three years and photos of it are unknown. It was just a single platform and closed in 1873 when the single line to Chingford opened.

Therefore, the first passenger trains on the Chingford branch to Shern Hall Street used the little-known Lea Bridge Curve and latterly Hall Farm Junction north of Lea Bridge station. The Great Eastern Railway thought that there would be little demand in 1870 for the new branch, and only a shuttle service was provided to and from Lea Bridge until the direct line across the marshes from Hackney Downs opened in 1872. Initially, the section from Hall Farm Junction to a temporary terminus at Walthamstow was just a single track, opening in the following year. Building continued towards Chingford with intermediate stations at Wood Street and Hale End (Highams Park), again just a single line. The GER installed a second line all the way to Chingford in 1875. The new double line caused a great upsurge in house building, as forecast by the GER. Bearing in mind that the majority of the line was quite rural in the 1870s, the Company's gamble certainly paid off. In 1870 the population of the Walthamstow area was just 11,000, but by 1900 it had rocketed to almost 100,000. Regular passenger trains over the Hall Farm Curve were an early casualty of the First World War, being dropped from the timetable in October 1914. Bank Holiday special trains occasionally used the curve, but these were curtailed by the outbreak of the Second World War in 1939.

The first Chingford terminus was a simple wooden building serving just one platform. Due to the popularity of Chingford and Epping Forest for day trippers, it quickly became inadequate and was replaced by an imposing double-storey brick built structure on a different alignment in 1878. Remarkably, the old station continued to stand, albeit rather derelict, until 1953 and the associated sidings were used as a coal depot until 1965. The new Chingford terminus had the appearance of a through station, due to the possibility of a future extension to High Beech on the edge of Epping Forest, but this never materialised. Locomotive servicing facilities were provided at Wood Street in 1879 with a two-road engine shed and a coaling stage. This closed when the line was electrified in 1961.

Once established in the late 1870s, the Chingford branch thrived and widespread housing development gradually radiated from the intermediate station's environs in the following decades. In 1880 the Company introduced a new Stratford to Chingford service over the Lea Bridge curve and Hall Farm Junction, but this was initially only one train every two hours! Between 1883 and 1888, Sunday excursion trains ran between Chingford and Fenchurch Street. By the time the First World War broke out in 1914, trains on the Liverpool Street to Chingford line were running at full capacity and overcrowding became a severe headache for the GER. Post-war, a plan was hatched to increase the train service by an astonishing 50% with signalling improvements and additional turnround facilities at Liverpool Street. Additionally, locomotive bays were constructed at the country end of the West Side platforms, which enabled the departing locos to be very quickly attached to the rear of an incoming service. In the peak hours, a train would arrive and depart every two minutes. The alterations took two years to complete and the new timetable was introduced in 1920. This equalled what other railway companies could only achieve with electrification, but the GER's method only cost £80,000 as

STRATFORD LNER POLYGON A train journey from Stratford to Chingford during the LNER period would have departed from Platform 12, the ramp of which is just in view on the left. To the right is the original Stratford Works, some of which dated back to the Northern & Eastern Railway days. The GER Main Works Office and Drawing Office building is to the right. This stood empty for decades and demolition took place in 2001. Beyond is the quirky Polygon signal box which straddled a track leading into the former roundhouse building, which has been demolished. A steel framework is just visible for a new structure.

opposed to an estimated £3 million for electrification; money that the Company could ill-afford. Along with the branch line from Hackney Downs to Enfield Town, the service gained the nickname *The Jazz*, due to the brightly coloured stripes underneath the coach gutters, denoting which class of travel each coach was intended for. The GER Chairman, Sir Henry Thornton and his Superintendent of Operation, F. V. Russell, were highly praised for what was widely acclaimed as being *The Last Word in Steam-Operated Suburban Train Services*. These were the most intensely worked steam hauled passenger services, not only in the UK, but the world. Further improvements were carried out on the Chingford line by the LNER, who invested in colour light signalling throughout. The engineering and installation work took two years and was completed by 1938. Train services ran normally between 1936 and 1938, which was a formidable achievement when compared to the disruption caused these days by engineering work.

The introduction of the ubiquitous A. J. Hill designed L77 (N7) 0-6-2T class by the GER in 1915 was a foretaste of what would become the most commonly used locomotives on *The Jazz* trains in North East London. The outbreak of WW1 caused a delay in the construction, meaning that only two locomotives were built in 1915, numbered 1000 and 1001. Post war, a further ten locomotives in the class emerged from Stratford Works during 1921. By 1924 another ten Stratford built engines were in service. The grouping in 1923 and creation of the LNER saw new opportunities for the Locomotive Operations Department to order more N7s, which were sanctioned by the Chief Mechanical Engineer H. N. Gresley who favoured their potential. These were built at Gorton, Doncaster, Robert

Stephenson's and William Beardmore's Works. Eventually there was a grand total of 134 N7s and Stratford had by far the largest allocation, with no less than 105 of them on call in the 1950s, mainly for NE London suburban traffic. The sole surviving N7 No. 69621 is currently undergoing a lengthy overhaul at the East Anglian Railway Museum, at Chappel & Wakes Colne in Essex. This was the last N7 built at Stratford Works, emerging in 1924. We look forward to seeing it back in steam with its distinctive thumping Westinghouse pump.

The GNR also introduced the Gresley designed Quad-Art and Quint-Art coaching stock for Kings Cross suburban services just before the outbreak of WW1, and these were gradually introduced on the Jazz service after the war. The articulated bogies gave a considerable weight saving and assisted the acceleration times on the Chingford and Enfield Town lines. A single Quad-Art rake would carry over 600 passengers, so their introduction solved much of the overcrowding. They were still being used up to the end of NE London steam-operated services in 1961 and just one set has been preserved. It has been beautifully restored in LNER teak livery by the North Norfolk Railway.

Curiously, when the Chingford line was being electrified in the early 1960s, the Hall Farm Junction to Lea Bridge curve had overhead wires installed. However, the planned overhead line (OHL) installation between Copper Mill Junction and Stratford did not take place, and the wires were removed circa 1965. No electric trains ever ran over the curve and the last movements were diesel hauled parcels trains to Hoe Street, ending in 1967. The Copper Mill Junction to Stratford OHL work was finally carried out in 1989. Better late than never!

By 1961 electric trains saw off the remaining steam hauled services and a new era began. That is a story for another day.

POLYGON GER DECAPOD 1903 Whilst we are at the Polygon engine shed, this gives me a perfect excuse to use a delightful glass plate negative which has never been published. The GER had an audacious plan to build a steam locomotive that could accelerate as fast as an electric train (or tram; their other great competitors). James Holden designed the mighty 0-10-0WT A55 *Decapod* locomotive seen here outside the Works Offices on the right, with the signal box in view on 12th January 1903. It weighed 80 tons and under test conditions could indeed accelerate a passenger train from a standstill to 30mph in 30 seconds. However, it was far too heavy and widespread development of the class would have necessitated the strengthening of many bridges, so just three years later it was withdrawn and converted to an 0-8-0 tender locomotive for freight work. It is only the bunker view, but the negative for the side view was rather out of my price range!

POLYGON BUILDINGS 1967 (1) The years after the end of steam in 1962 have not been kind to the former Stratford Locomotive Works and Polygon sidings. This February 1968 scene is taken from virtually the same viewpoint as the previous image. The tracks have been filled in and the buildings lie empty, waiting for demolition. Only the Works Offices escaped the 1960s bulldozers. The corrugated iron building which was being constructed on the site of the Polygon roundhouse in the previous view had a very short life. On the left is Stratford Old Yard, used mainly for coaches and parcels vans. *Photo: Roy Lingham.*

POLYGON BUILDINGS 1967 (2) Ex-Stratford driver Roy Lingham had the foresight to photograph the dying days of the old locomotive works in 1968 shortly before the majority of the buildings were razed to the ground. His collection contains about 20 views and could form an article on their own merit. One of the only surviving artefacts still in situ from this former bastion of steam locomotive engineering, is a small section of brick wall in Leyton Road opposite The Railway Tavern, which has a long association with footplate crews, being the venue for some of the very first ASLEF Union meetings in the 1880s. A tradition which survives to this day with an Annual Stratford Reunion every April. *Photo: Roy Lingham.*

CHOBHAM FARM JUNCTION SIGNAL BOX GER This view from the final days of the GER in 1922 is looking towards Stratford station from Chobham Farm Junction. To the left are the Carriage Workshops which covered a sprawling area containing sawmills, carpenters, furniture makers and paint shops. The vast majority of coaching stock had wooden bodies at this time constructed from high quality teak. Thankfully, some have been beautifully preserved. The junction to the right entered that Holy Grail; the vast iconic Stratford steam engine sheds. Behind the photographer was Temple Mills East Junction.

LEA JUNCTION BOX 1921 Anybody who is familiar with Stratford may recall that there were about a dozen signal boxes in the area, controlling hundreds of signals prior to the 1948 resignalling scheme when the majority were closed, including Lea Junction seen here. This formed part of the triangular junction on the route towards Victoria Park and the North London line. The three-sided junction is still in regular use. This rare image was captured in 1921 during the GER period. We are looking towards Stratford and the River Lea is in the foreground. Substantial drainage work is being carried out by the engineers, and a large stockpile of new pipes is seen in front of the fence along Carpenters Road.

TEMPLE MILLS EAST A view from Temple Mills Lane overbridge. A great vantage point to witness traffic going in and out of Temple Mills Marshalling Yard. I stood at this very spot for hours on end in the early 1970s. Sadly, steam had departed a decade earlier and the only steam locomotive I saw here was *Flying Scotsman* in 1984 when it headed The Queen Mother's special to North Woolwich for the museum opening ceremony. This image shows the now preserved N7 No. 69621 working The East London Suburban Railtour No. 2; a special organised by The Railway Correspondence and Travel Society (RCTS) on 28th April 1962, (details on the Six Bells Junction website.) On this leg of the tour, the train had departed Palace Gates, then passed through Stratford before returning to this spot via Channelsea Junction. It then headed in the other direction, becoming a very rare 1960s passenger working over Hall Farm Curve on its way to Chingford.

LEYTON ENGINEERS CRANE Just to the right in the previous image was the huge Leyton Engineers Depot, which was home to this self-propelled Taylor & Hubbard Ltd. crane captured in the 1940s. The yard was used by the Permanent Way Department and Signalling Engineers. I had the privilege of working trains in and out of this yard; in addition, I spent many shifts on the 08 diesel pilot. A large section of the yard was used to store ballast, and track engineers would often assemble new pointwork to check that everything was correct before disassembling the jigsaw of components, which were then loaded onto wagons for the work site.

TEMPLE MILLS DECEMBER 1958 Temple Mills Marshalling Yard is just three months old when this image was captured from the top of the Temple Mills Wagon Shop's power station in December 1958. After a £3 million pound investment programme, this vast yard comprising some 50 sidings, were controlled from the tower which is just visible in the distance. Wagons pushed over the hump were allowed to run into their chosen siding under gravity, with rail mounted retarders used to slow their descent into each road. On the right is an ex-WD 2-8-0 locomotive and an almost new Brush Type 2 'Toffee Apple' in the D5500-D5519 series waits to depart on the left. The two 08 diesel shunters were kept very busy 24 hours a day.

LEA BRIDGE 1975 The reception roads for Temple Mills Marshalling Yard's hump can be glimpsed in this view of a run down Lea Bridge station in 1975. Passenger services here were gradually eroded by a lack of patronage in the 1970s and by the time of closure in July 1985 there were just a handful of DMU services between Stratford and Tottenham Hale in the rush hour only. A delightful nostalgic BR dark blue enamel totem sign clings to an LNER lamp post; reminders of a past life and more prosperous days for this neglected route. After four decades of lying in ruins, the station came back to life in 2016 as an £11 million new station. *Photo: Alan Young*

LEA BRIDGE PARCELS DEPOT This is an early 1940s view of Lea Bridge Parcels Depot under construction. To the left is the once-elaborate original station building designed by the architect Sancton Wood. During WW2 it was heavily damaged and the roof was demolished for safety reasons. The remains of the structure were removed in 1960 only to be replaced by a rather austere and uninviting ticket hall seen in the previous view. The parcels depot survived until the late 1970s with occasional traffic. The building was still standing when the last passenger train departed in 1986, but was subsequently removed to make way for a new road.

Opposite top: **HALL FARM JUNCTION BOX** This is the rarely photographed Hall Farm Junction, showing the 1873-built GER signal box as it appeared in 1911. The tracks leading to the right joined Lea Bridge Junction (the box closed in April 1958, but the junction survived for another decade); another rarely photographed location. To the left are the tracks from Copper Mill Junction (added in 1885), whilst those passing underneath the photographer, who was on top of a signal post, are those from Clapton Junction. The body of water is Walthamstow Reservoir and St. James Street station is out of view in the distance. The signal box was abolished and replaced by a ground frame during the LNER 1935-1938 colour light signalling programme.

Below: **HALL FARM JUNCTION MAP 1915** This 1915 GER map shows the Walthamstow and Lea Bridge Curves meeting at Hall Farm Junction, top right. Copper Mill Junction is just off the map on the top left hand corner. Clapton Junction is at the bottom and Lea Bridge Junction is out of view on the right. Regular passenger services ceased over the Walthamstow curve on 6th September 1926.

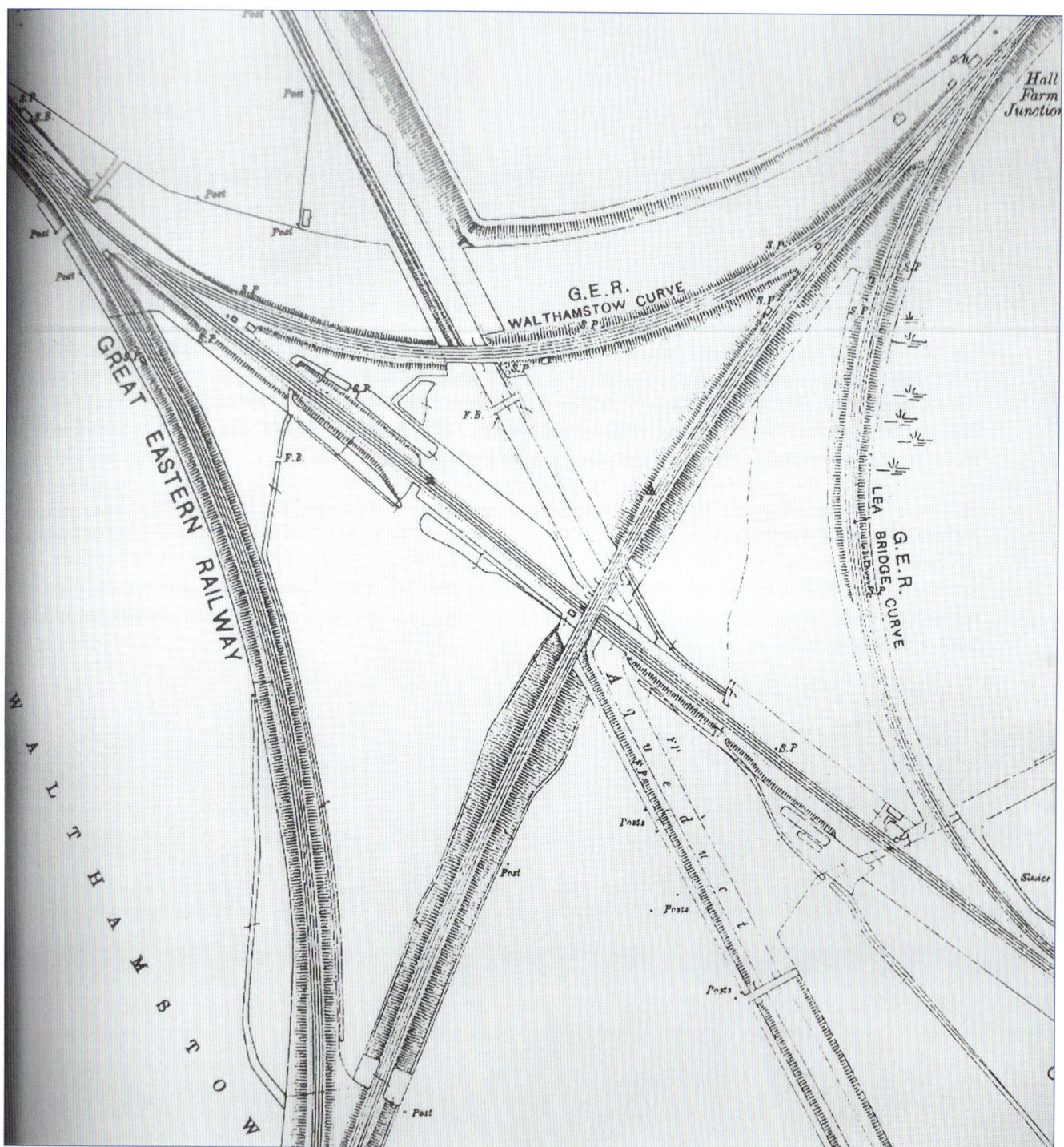

HALL FARM JUNCTION MAP LNER

This map is reproduced from the LNER 1935 resignalling plan and clearly shows the positions of the three signal boxes and the track layout north of Lea Bridge Junction. The use of trailing spring points at Hall Farm Junction is also noteworthy.

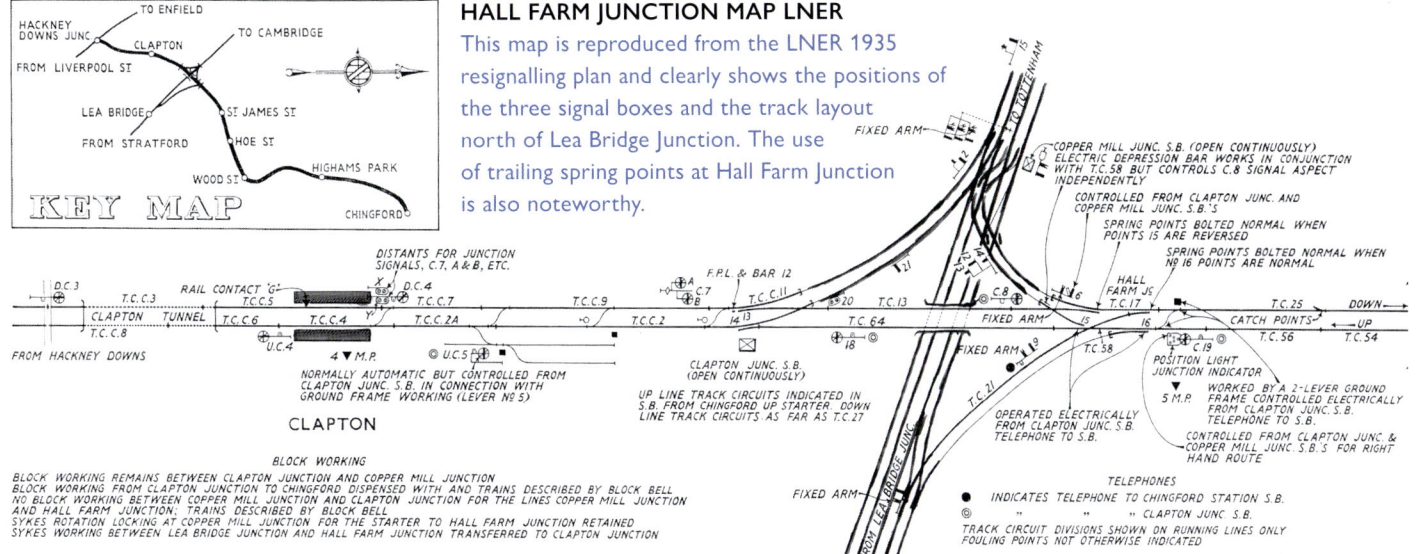

COPPER MILL JUNCTION BOX Copper Mill Junction is very much alive today, with very frequent electric services heading along the Lea Valley from both Liverpool Street and Stratford. Here, we are transported back to 1911 when a third pair of tracks head off towards Chingford and Hall Farm Junction on the right. This link, known as the Walthamstow curve, was severed in 1960 and lifted shortly afterwards. The GER Type 7 signal box was moved to the right in the early 1930s to accommodate two additional freight tracks, and unconnected groundworks appear to be underway here.

COPPER MILL JUNCTION D5512 A rare colour view of Brush Type 2 'Toffee Apple' A1A-A1A No. D5512 partly obscuring Copper Mill Junction signal box in June 1964. Closure of the box came in February 1969 when all the signalling was superseded by Hackney Downs and Temple Mills West boxes. This early batch of just 20 locomotives was only ever allocated to Stratford, and having the electro-magnetic (Red Circle) control system made them incompatible with the rest of the class, which had the electro-pneumatic (Blue Star) control system. The slow decline in freight during the 1970s saw the whole class withdrawn by October 1980 and only D5500 (31018) was saved for the National Collection.

CLAPTON JUNCTION BOX
To complete the trio of signal box images shown on the accompanying map, this is Clapton Junction box as it appeared in 1959, shortly before the overhead wires were installed for the Chingford line. During WW2 the box suffered bomb damage, and a brick base was added to the lower half to stop it from falling into the Hackney Marshes! Closure came on 22nd May 1960 during resignalling and electrification work and its duties were taken over by the new Hackney Downs box. In the distance can be seen the various industries that surrounded Lea Bridge station.

GER JAZZ COACHES
The 'Jazz' is a term often associated with the North East London suburban workings, but the term was first used in America during 1917 when the *Original Dixieland Jazz Band* captured the public's imagination with a lively new music style. The culture spread across the Atlantic to our shores and became popular here, with Jazz clubs popping up in London and then spreading across the country. 'Jazzy' colours, as they were termed, was a phrase adopted by the media to describe the bright colours applied to the GER suburban coaches, denoting the class of travel. First Class compartments had a yellow band, seen here on the right at Liverpool Street and Second Class compartments had a blue band above.

ST. JAMES STREET The first station on the Chingford branch is St. James Street seen here in the final days of steam, just before the full electric service started in the final months of 1961. The familiar traction is of course an N7, No. 69670, hauling a five-coach articulated set of Gresley coaches on a Liverpool Street to Chingford service. The 1878 GER 18-lever frame signal box was sited at the country end of the Down platform, but this was abolished during the LNER 1938 colour light resignalling programme and photos of it are scarce. *Photo: Tony Wright*

HOE STREET BOX GER
A rare GER view of Hoe Street station in Walthamstow as it appeared circa 1910. We are looking towards Chingford and a thriving goods yard is full of pre-grouping and private owner wagons. F. Warren's coal sidings are rammed with wagonloads of domestic coal. On the left are sidings serving a timber yard. The original 1873 built 22-lever signal box seen here was sacrificed in January 1938 during the LNER colour light resignalling project. Just a short walk from here is the Tottenham & Forest Gate Junction line's Walthamstow (Queen's Road) station.

HOE STREET STATION N7 0-6-2T No. 69646 pauses briefly at Hoe Street in 1961 with a Chingford bound service and the overhead wires indicate the locomotive's impending doom! Vast numbers of N7s met their grisly end at Stratford during the early 1960s. The station was renamed Walthamstow Central in 1968 and became an important interchange upon the opening of the new London Underground Victoria line. Out of view behind the Up side platform was a parcels depot, and diesel hauled trains still served this until 1967. *Photo: Tony Wright*

WOOD STREET GER
Wood Street's two-road engine shed opened in 1879 and was lengthened to accommodate eight tank engines by 1900. The view here dates from 1911 as we look towards London. The signal box is an early 1900s GER Type 7 38-lever replacement for the original 1878 structure. It closed in June 1960. Conditions for coaling and servicing engines were very basic here when compared to the colossal Stratford depot. In the 1930s the LNER demolished the coal stage due to its deteriorating condition and this was never replaced. Locos were subsequently coaled by hand from adjacent wagons – harsh work. The end of steam saw most of the drivers and firemen transfer to Chingford depot. Not all of them took to EMU work and some left the railway altogether.

Above: **HIGHAMS PARK** A wonderful GER view of Highams Park & Hale End as it appeared in 1911. The nostalgia oozes from this panorama and the original 1878 signal box located on the Up road platform makes this a rare image. Note the covered subway, built in 1909, which enabled passengers to catch their trains even when the level crossing was closed. A busy two-road coal yard is on the right. Enamel advertising signs and gas lighting abound. The station opened in 1873 as Hale End and was renamed Highams Park & Hale End in 1899. The Hale End suffix was dropped in 1969. A larger signal box was built by the LNER in 1925 on the London side of the level crossing, and this was still in use until 2002, although it was only used to control the level crossing after the 1938 colour light signalling was installed. It still stands, 100 years after it was built.

Opposite: **CHINGFORD 9th March 1957** Our journey's end and the layout of Chingford station as a potential through station is clear. The hoped for extension to Epping Forest (High Beech) was just a dream. Queen Victoria arrived by train at Chingford in 1882 and publicly declared that Epping Forest would be for the people for all time. Platform 4 on the right closed decades ago, but the 1878 double storey building stands in a very well-preserved condition today. Nearly 2 million passengers now use the station each year. This pre-electrification view was taken on the 9th March 1957 looking towards the buffer stops. The 1920 built GER Type 7 signal box with a 60-lever frame was a replacement for the original 1878 structure. It closed in 2002. Note the short spur behind the box; a remnant of the Jazz service, where a departing loco would be waiting to move onto an arriving train within minutes. *Photo: Andy Grimmett Collection.*

Author's Note: I would like to express my gratitude to Andy Grimmett for his considerable help with regard to signal box matters, which is not my strongest subject. Many publications have been consulted; in particular the Middleton Press books by Jim Connor and H. V. Borley's 'Chronology of London's Railways' published by The Railway & Canal Historical Society in 1982.

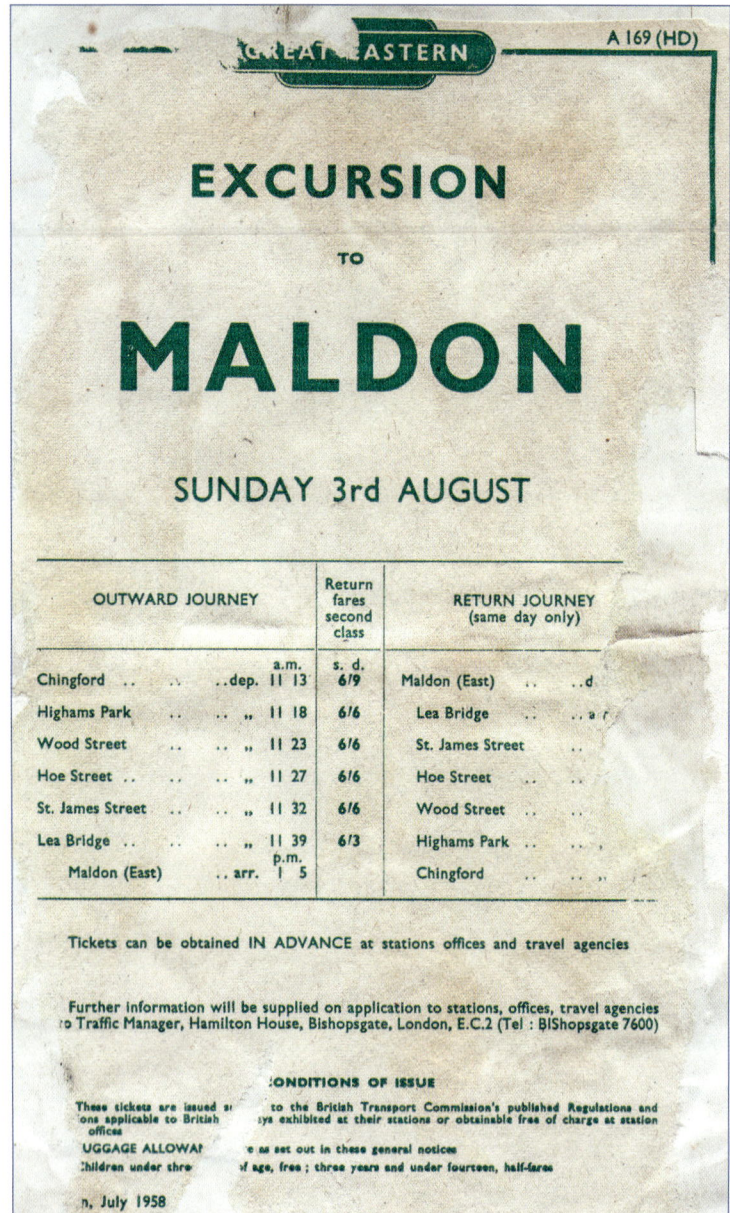

All Photos: Gordon Wells

CHINGFORD N7/1 0-6-2T No. 69645 stands at the buffer stops in the carriage sidings at Chingford in the early 1960s. Alongside is the bunker of a classmate in platform 1. This engine spent most of its career at Stratford working suburban trains in and out of Liverpool Street. One of a batch built at Gorton Works in Manchester by the LNER between 1925 and 1926, it was recorded by long time Stratford driver Gordon Wells when he was a young fireman on *The Jazz*. His driver smiles for the camera and the job satisfaction is all too evident. Gordon was a true gentleman and hugely passionate about steam in all its many forms. He owned the Manning Wardle 0-6-0 steam locomotive *Newcastle* for several decades. This is now at Beamish.

CHINGFORD Chingford hosts N7/2 0-6-2T No. 69677 at the head of a service to Liverpool Street in the early 1960s. This engine spent time allocated to Parkeston (Harwich) 30F depot in the 1950s, but by the end of the decade it had returned to Stratford, spending its remaining few years on *The Jazz* and North Woolwich to Palace Gates services. The poor old Westinghouse brake pump is visible. These were often beaten by the fireman with a hammer if they packed up in service and it was a common sight to see them covered in dents. Note the old enamel tea can in the cab. The No.1 priority for all footplate men has always traditionally been a cup of tea for the journey ahead! Some things never change, but an N7 on the Chingford line will never be seen again.

BR HANDBILL MALDON This delightful 1958 handbill advertises a Sunday Excursion from Chingford to Maldon travelling over Hall Farm Curve. The timing between St. James Street and Lea Bridge is just three minutes! Today, the same journey would have to be via Hackney and Stratford, taking at least 45 minutes, it is quicker to walk. There have been many requests from commuters and rail pressure groups to have Hall Farm Junction reinstated. As recently as 2023, the Greater London Authority announced its backing for such a scheme, but it will not happen for many years, if at all.

IN DEFENCE OF SIR THOMAS BOUCH

BY IAN LAMB

As we rightly celebrate the bi-centenary of the Stockton & Darlington Railway, it is worth remembering that for four years Thomas Bouch was actually involved in that company's progress and development, especially that of the Stainmore route over the Pennine hills.

Bouch was born in Cumberland, and eventually lived in Edinburgh, working as a railway engineer. Perhaps the uniqueness of the man was when – as manager of the Edinburgh & Northern Railway – he introduced the first roll-on/roll-off train service in the world, between Burntisland and Granton across the Forth estuary in 1850, but for me it was always his incredible Belah viaduct on the former LNER's Stainmore line over the Pennines that showed his greatness.

Anyone who knows the Stainmore route across the backbone of England will be only too aware of how challenging it was to survey and construct a railway across this remote and desolate terrain. However, all of Bouch's bridges stood the test of time and weather until being physically dismantled in the 1960s, well over a century later.

John Yellowlees records that the genius of the man perhaps came from the genes of his seafaring father when he devised the 'Floating Railway', which comprised three main elements. There is an inclined pier at the dockside, a flying bridge to the vessel's deck and a flat deck on board the boat, lined with rails. A moveable framework rolls up and down the pier on 24 wheels to suit the state of the tide, the hinged linkspan being operated by a steam winch. This became the outline for almost all subsequent train and car ferries for the next 150 years, by which time the era of train-ferries were almost over, and 'roll-on/roll-off' is now the norm for road vehicles. The first vessel was duly named *Leviathan*.

Wikipedia is credited with the following information. Over a total distance of 5 miles (8km), the 'floating railway' carried only freight traffic, there being separate passenger ferries. Three (later five) 'goods boats', as they became known, carried the wagons – with a maximum capacity of between 20 and 40 depending on the vessel – across the Firth of Forth until the opening of the Forth rail bridge in 1890. In the first six months alone, over twenty-nine thousand wagons had been transported. No doubt Bouch was rightly satisfied with his maritime achievement, and therefore looking forward to what the Tay estuary could bring.

As a 17 year old he began his civil engineering career as an engineering assistant on the Lancaster & Carlisle Railway, before spending a year in Leeds in 1844, and then for four years took on the role of a resident engineer with the

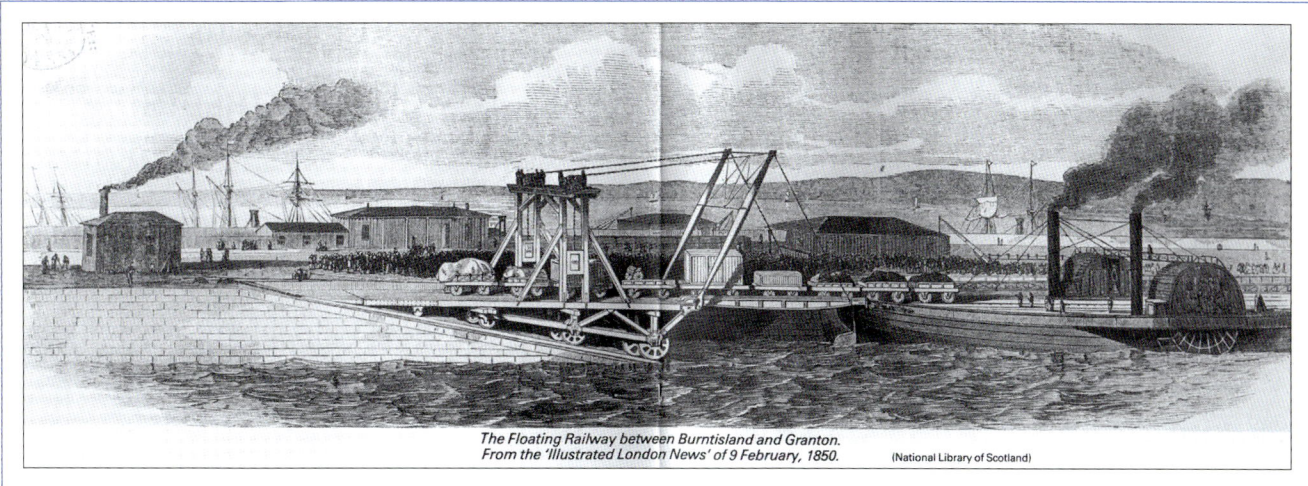

The Floating Railway between Burntisland and Granton. From the 'Illustrated London News' of 9 February, 1850. (National Library of Scotland)

Surely one of the finest views from a railway anywhere in Britain; the approach to Dundee across the Tay estuary to the distant Sidlaw hills. Gresley A3 Pacific No. 60052 *Prince Palatine* builds up speed on the sharp curve through Wormit with a southbound heavy freight. Note how calm the Tay estuary is, but be reminded of the fate that fell on the original bridge from an extraordinary storm on 28th December 1879. *Photo: Sandy Murdoch © Transport Treasury*

1937 • Banked from the rear by an unidentified loco Worsdell Class J21 0-6-0 No. 1564 heads a passenger train over the magnificent Belah Viaduct. *Photo: Neville Stead Collection © Transport Treasury*

Stockton & Darlington Railway, which preceded the young innovative engineer's appointment with the Edinburgh & Northern Railway (precursor to the North British Railway) in January 1849 as their manager.

Nevertheless I wonder what he was thinking in the mid-1870s when he first saw the magnificent view to Dundee across the very wide River Tay. No doubt pausing for a moment – or probably a lot longer – to fully take in the fact that he was expected to build a bridge that would safely cross this expanse of water? Sadly, such a project would eventually be his downfall.

When the North British Railway merged with the Edinburgh & Glasgow Railway in 1865, they appointed Stephenson and Toner to design a bridge for the Forth, but instead the commission was given to Bouch around six months later. Shudder the thought, but – like the original Tay Bridge – it would have been single track!

On that tragic night of 28th December 1879, during a violent storm, the relatively new Tay Bridge collapsed as the 4.15pm train from Edinburgh was crossing. There were no survivors from the 75 passengers and crew. Elements of the supernatural were recorded by a farmer in East Lothian who said to his friend on leaving church, "I have just seen the Tay Bridge go down". Many devout Christians were not slow to say it was due to trains running on Sundays.

It is said that if something doesn't look right, it probably isn't; and that's my view regarding lots of 'flimsy' structures. Compared with Bouch's Belah viaduct – which was double-track and looked immensely strong (though throughout the latter period of its existence – as locomotives got heavier and heavier – only one train was allowed to cross at any one time) – the original Tay Bridge does not look as if it could withstand any wind, never mind storms. I'm also reminded of that old saying, "Ships are safe in harbour, but that's not what they were designed for". However, strengthening work was carried out on the Belah viaduct in 1955 to enable BR Standard Class 4 locomotives to cross.

The public inquiry into the ill-fated Tay Bridge disaster found it to be "badly designed, badly constructed and badly maintained," with Bouch being "mainly to blame" for the defects in construction and maintenance and "entirely responsible" for the design defects. Whilst appreciating that on reflection the tragedy happened during high winds for which he had not properly accounted, the Board of Trade thereafter imposed a lateral wind allowance of 56 lbs/ft^2. His 1871 design had taken a much lower figure of 10 lbs/ft^2 on the advice of the Astronomer Royal, although contemporary analysis showed it would likely have stood, but the engineers making the analysis stated that "we do not commit ourselves to an opinion that it is the best possible design."

It is recorded in Aberdeen University files that the last men to see the ill-fated North British Cowlairs engine No. 224 and its carriages were the signal box staff on the south side of the river, when the single-line tablet was handed to the fireman. Duty indeed came first, and one can easily imagine what it was like when the locomotive accelerated on to the straight track of the original Tay Bridge in that howling blast. Soon the train entered the high girders, presenting the wind with a solid objective of which to exert its appalling force. Once on the high girders the train plummeted to the river bed.

Total confusion reigned. People on both sides of the estuary reported seeing lights plunging off the bridge, but not even the signalmen on either side knew what had happened for some considerable time. The actual tragedy was bad enough but first garbled reports – emitted in panic from the North British head office in Edinburgh – put the death toll at 300, well beyond the carrying capacity of the train.

On a personal front I have experienced what extraordinary foul weather can be like when in my late twenties leading a group of walkers on Ben Nevis. We were on 'all fours' half way up Ben Nevis before succumbing to the strongest wind that I ever encountered, so we gave up that venture and headed down to safety.

To all intents and purposes it could be argued that Thomas Bouch built many of his structures 'on the cheap'; but in his defence, no doubt he would state that his works were all of a light and inexpensive character, and if he provided a first-class railway – "one upon which any speed attainable by a locomotive engine could be run with perfect safety and ease without any extravagance – then he should only have done his duty, but if he failed then he should deserve all the strong public condemnation and discredit attaching to the failure of light works". Perhaps these comments would come back to haunt him, and ensure that the Forth Bridge was constructed safely far beyond what it needed to be.

Summer 1897 • Views from the North (above) and South of Bouch's original Tay Bridge. The photo below shows the Tayport line branching off to the right. *All photos on pages 24 and 25: National Library of Scotland. Creative Commons Attribution 4.0 International Licence*

Last standing Pier North end of Gap

Images of destroyed piers and girders of Tay Bridge, wreckage of the steam engine and carriages salvaged from the River Tay. The bridge collapsed because of high winds on 28th December 1879. Commissioned by John Trayner on behalf of the Board of Trade.

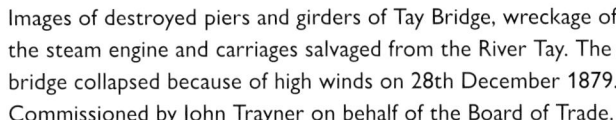

A view taken by distinguished Victorian photographer, George Washington Wilson, of the new Tay Bridge standing next to the remains of the old. Designed by William Arrol, who was already at work on the Forth Bridge, it officially opened on 13th June 1887. Wormit station is in the foreground and serves the Tayport line and its ferry services to Broughty Ferry Pier.
Photo: University of Aberdeen. Creative Commons Attribution 4.0 International Licence

Construction of Bouch's wrought and cast iron Belah viaduct commenced in November 1857 when Henry Pease, the new Member of Parliament for South Durham, laid the foundation stone. Costing £31,630, contractors Gilkes, Wilson & Co erected Belah in a little over two years. At 196 feet, it was the tallest bridge in England. The image shows Belah viaduct meeting its end after 102 years. *Photo: Hocquard Collection © Transport Treasury*

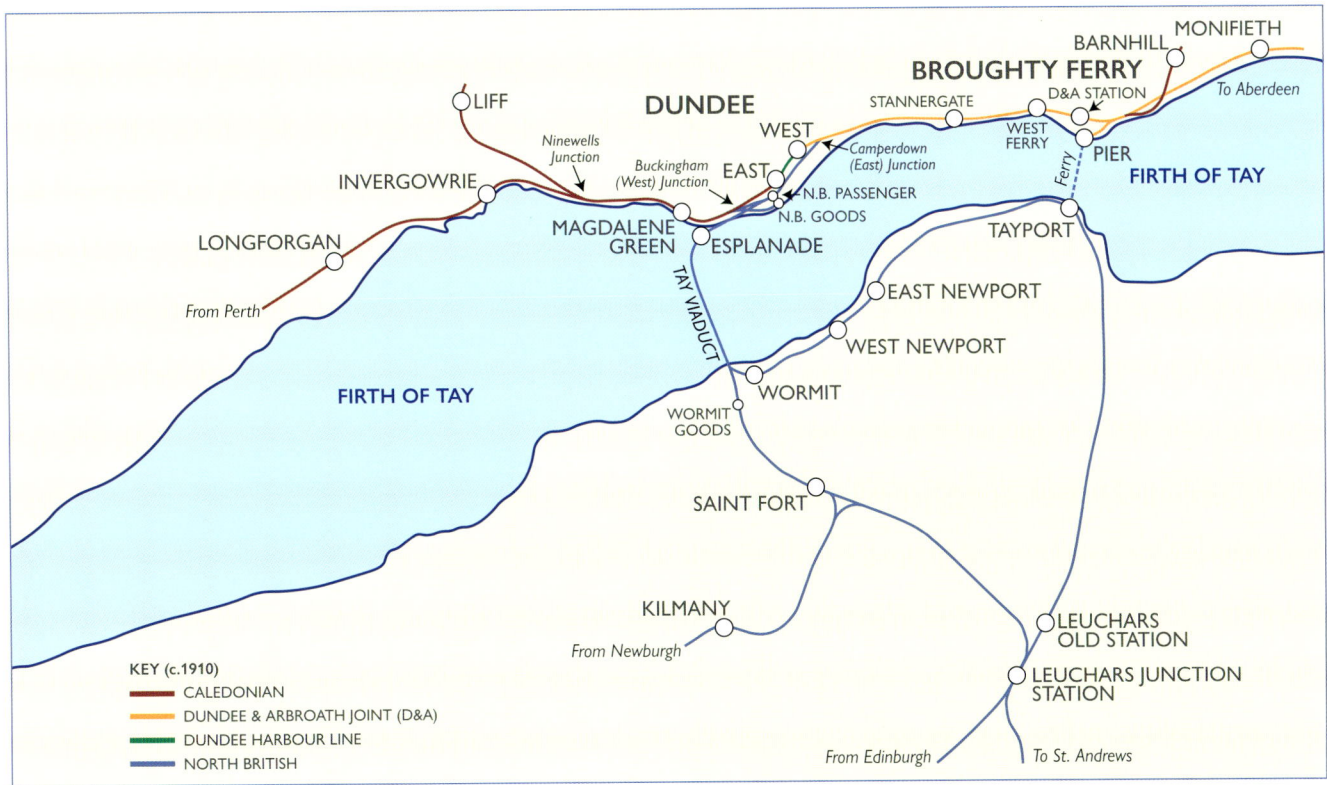

Among his many railway achievements was the Eden Valley Railway which ran from Kirkby Stephen to Penrith. In response to a toast during a dinner after the cutting of that railway's first sod, gave some insight to his bridge building philosophy, making considerable use of lattice girder structure on conventional masonry piers. They were considered as one of the lightest and cheapest ever erected.

He had the ability to construct branch lines at a capital cost that might allow them to pay their way, especially if operated frugally. On the general railway itself, in 1854 he advised the directors of the Peebles, St. Andrews and Leven branches to work the lines themselves, believing that they could do so much more economically than a larger company. Nevertheless, eventually they all came under the jurisdiction of the North British Railway.

The official opening of the first Tay Bridge took place in May 1878. So impressed was Queen Victoria after crossing the structure during late June in 1879, that she awarded Bouch with a knighthood.

In Andre Gren's book *The Bridge is Down* he states, "While the blame for the Tay Bridge Disaster might have been appropriately laid at Bouch's door, the Government's system of inspection also gravely failed the men, women and children who lost their lives from the tragic circumstances on the night of 28th December 1879."

4th September 2024 • Due respect to a great railway engineer buried in Dean Cemetery, Edinburgh. *Photo: Ian Lamb*

Within a year Sir Thomas Bouch was dead as a result of nervous strain according to the doctors; a broken heart was the belief of his family and friends.

NEVILLE ON THE SOUTH BANK

BY PAUL KING

I first came across Neville Stead when I was purchasing photographs for the fledgling Grimsby-Louth Railway Preservation Society, now the Lincolnshire Wolds Railway. Immediately we struck up a friendship that lasted through to his passing.

Amazingly, we never met, although I missed him by minutes at Coalville open day in 1983. His collection was invaluable when writing my first book – *Railways Around Grimsby*, P. K. King and D. R. Hewins, Foxline Publishing, 1988 – and I have used photographs from his collection in my subsequent books and articles. We would often talk on the phone for an hour, but never more as charges applied. Over the years he amassed an archive of railway photographs second to none for a personal collection. It is from this archive that I have chosen the photographs used in this article.

Neville was a Hull lad, born and bred, and fiercely proud of his heritage, even though in later life, he lived in the north east. Occasionally, he ventured south of the Humber and, graciously, we allowed him entry even though he didn't have a passport. (Authors note: just like in many other areas, there is a fierce, yet friendly, rivalry between the two banks, we call the north bank, Hull, the dark side, they call us, Grimsby, cod heads.) Our willingness to allow him access to Lincolnshire's sacred land has left us with a legacy of photographs that both Neville and us Yellow Bellies (nickname of people from Lincolnshire) can be justly proud of.

In this two-part article, the first part will cover LNER days through photographs acquired by Neville. The second part will show mostly his own work and cover the BR era.

All photographs are from the Neville Stead Collection, many being now part of the Transport Treasury archive.

Left: The LNER were renowned for finding horses for different courses. Those that immediately spring to mind are the GER B12 4-6-0s sent north to the ex-GNoSR and the GNR K2 2-6-0s to the West Highland and D1s to other parts of the NBR. The GER Class J67/J69 found their way all over the system and there were several other moves, either temporary or permanent.

A familiar sight just a couple of miles to the north, at Hull Dairycoates, a Class J21 caused quite a stir when it arrived in Lincolnshire in the middle of 1931. After a short stay at Lincoln, No. 289 arrived at New Holland at the beginning of 1932, where it is seen later in the year alongside the rudimentary coal stage. It was soon transferred to Immingham and shortly afterwards found itself in East Anglia, where it remained until withdrawn from New England in 1937. Several members of the class made the move south but none stayed for more than a handful of years.

Above: After working the GNR suburban services in West Yorkshire and out of Kings Cross for a number of years, the C12 4-4-2Ts found pastures new across both ex-GNR lines and further afield. Over the years, several found a new home at Louth, where they were ideal for the lightly loaded local services to Mablethorpe, Bardney and Grimsby. There were already three on the books at the time of the grouping in 1923 and the last, No. 67398, left when the depot closed in December 1956.

No. 4013, seen here running around its train at the north end of Louth station, transferred from Copley Hill in June 1937 and remained at the depot until July 1953, when it left for Hull (Botanic Gardens) as BR 67352. The fourth member of the class to enter service, in May 1898, it was amongst the last to be withdrawn, in November 1958. Shortly before its demise, it was spruced up and appeared in BR lined livery, complete with the later BR logo, the only member of the class to be so adorned. A prime candidate for preservation in that condition? Unfortunately not, it was broken up shortly after withdrawal.

EASTERN TIMES • ISSUE 8

Opposite top: I recall Neville relating an interesting story about this photograph. He received a batch of negatives from someone that included a couple of reels of undeveloped film. He could see that they were old and they were that tightly wrapped he doubted he could do anything with them. He managed to unwind the film and develop it and this picture was amongst those on the film.

It shows Ivatt Large Atlantic, Class C1, No. 4418 at the south end of Louth station waiting time with an express for Peterborough and Kings Cross in the 1930's. Note the driver, leaning out of the cab waiting for the guard to give him the right away.

Opposite bottom: Not all shunting required the use of a locomotive, particularly in the more rural areas. Horses were often to be found manoeuvring the odd wagon around the yards at these smaller locations and Louth was no exception. It retained a shunting horse into BR days. In this 1930's scene, the Louth shunter is waiting the passing of a C12 tank, just visible under the station roof. It's home, or shed, was the van parked in the bay. Although he never said as much, all three of these photographs may well have come from the same reel of film.

Below: Aby, for Claythorpe, was one of the stations between Louth and Alford on the East Lincs line, some 7½ miles south of Louth. There are a few photographs taken of the station but few showing a train. Other than this one, I know of only one other, a WD 2-8-0 on a freight. An ex-GCR Class 8, LNER Class B5, 'Fish' engine, No. 5181, is seen heading north towards Louth and Grimsby with a morning stopping train from Peterborough in the late 1930s.

These locomotives, designed by J. G. Robinson, were primarily used on the express fish trains that left Grimsby in the late afternoon and early evening, hence the sobriquet 'Fish engine'. Eventually replaced by more modern designs, they transferred to other duties around the ex-GCR system and beyond.

Once the pride of the GCR, Robinson's Class 9P 4-6-0s earned a reputation for being heavy coal burners, even though they were strong and fast. The LNER even tried them on such trains as the Yorkshire Pullman after the grouping. When new, these locos, in their GCR livery looked magnificent but they never stayed anywhere for long periods. Later in their careers, four out of the six members of the class, were rebuilt with Caprotti valve gear in a vain attempt to improve their appetite for coal. This destroyed their looks and made for the ungainly locomotive seen below. The only one to survive nationalisation was the former No. 6166 *Earl Haig* which was substantially rebuilt with a B1 type boiler and other improvements in 1944. Unfortunately, it suffered from cracked frames and only lasted until 1949. The most famous of this class was the war memorial engine, No. 1165 *Valour*. It was withdrawn, as LNER No. 1496, from Lincoln on the last day of the LNER, 31st December 1947. There are few locomotives I can think of that were better candidates for preservation, but it was not to be.

No. 1498, formerly named *Lloyd George*, is seen in the final few months prior to withdrawal in December 1947, with a stopping train for Cleethorpes. It is approaching Brocklesby on the up fast line.

Above: A total of 94 of H. A. Ivatt's large boilered Atlantics were built. Classified C1 by the LNER, there was very little variation to the design over the years, although there were exceptions. No. 292, built as a four-cylinder compound, withdrawn 1927. No. 1300, built as a De Glehn compound locomotive, withdrawn 1924. No. 1421, built as a compound but rebuilt in 1920 to the standard design, withdrawn 1947. No. 279, rebuilt as a four-cylinder locomotive, rebuilt to two-cylinder, although non-standard with the remainder of the class, in 1937, withdrawn 1948. No. 1419, fitted with a booster to the trailing carrying wheels in 1923, removed 1935, withdrawn 1948. Alongside these major changes to design, the cosmetic changes made to No. 2877, the subject of the photograph were minor. As LNER No. 4447, it was rebuilt to the NBR loading gauge for trials between Newcastle and Edinburgh. The lower profile cab fitted by the GNR didn't provide sufficient side clearance and a higher profile cab with narrower sides was fitted. It retained this cab through to withdrawal in November 1949. After withdrawal, the locomotive survived for a further three years as a stationary boiler at Doncaster's Wagon Works and later at the Carriage Works.

Before moving to Sheffield on 3rd July 1949, incidentally, the day before I was born, it was based at Doncaster. It is seen here shortly after leaving Brocklesby in 1947, with a stopping train for Doncaster.

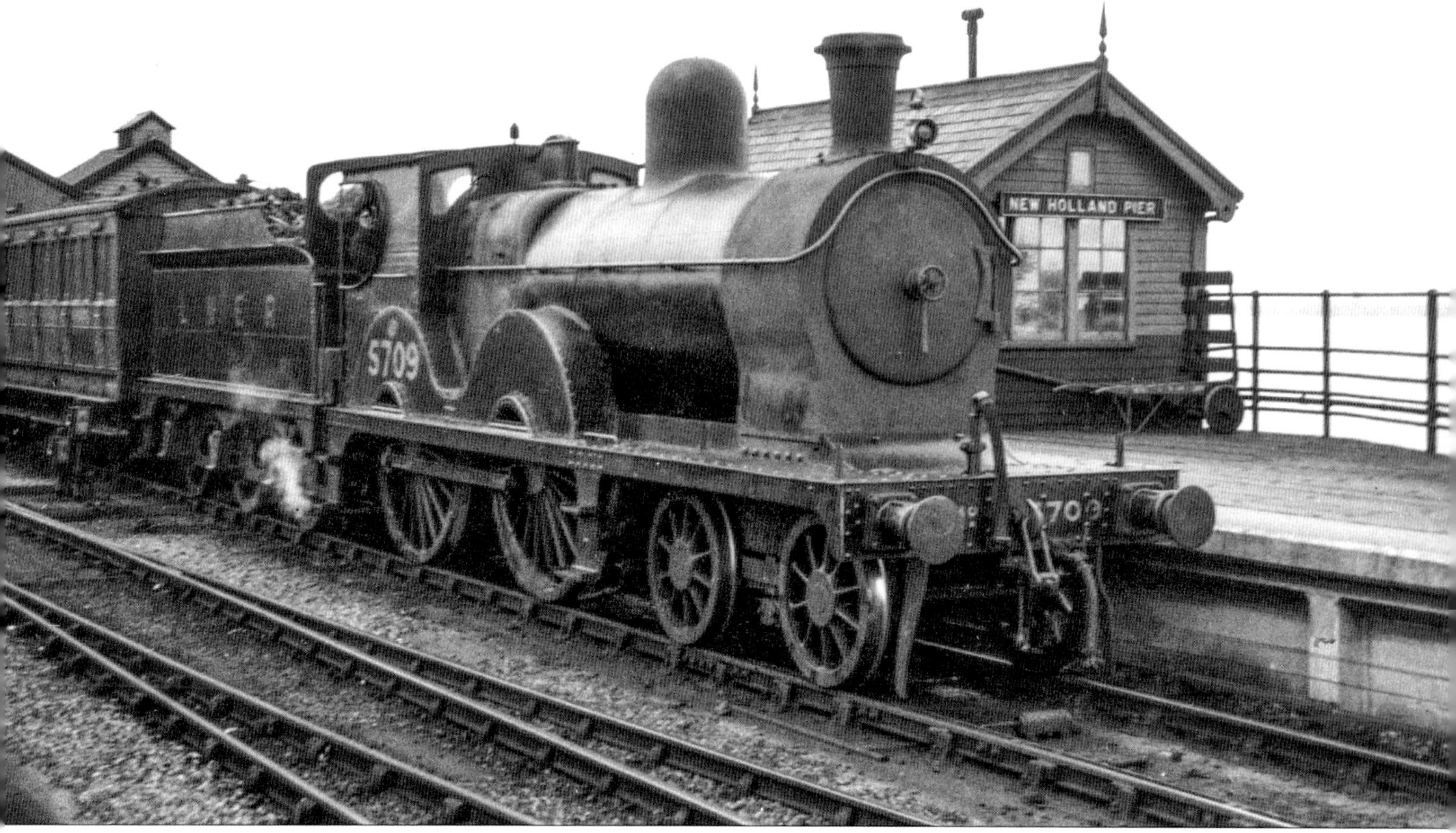
NEW HOLLAND PIER
L N E R
5709

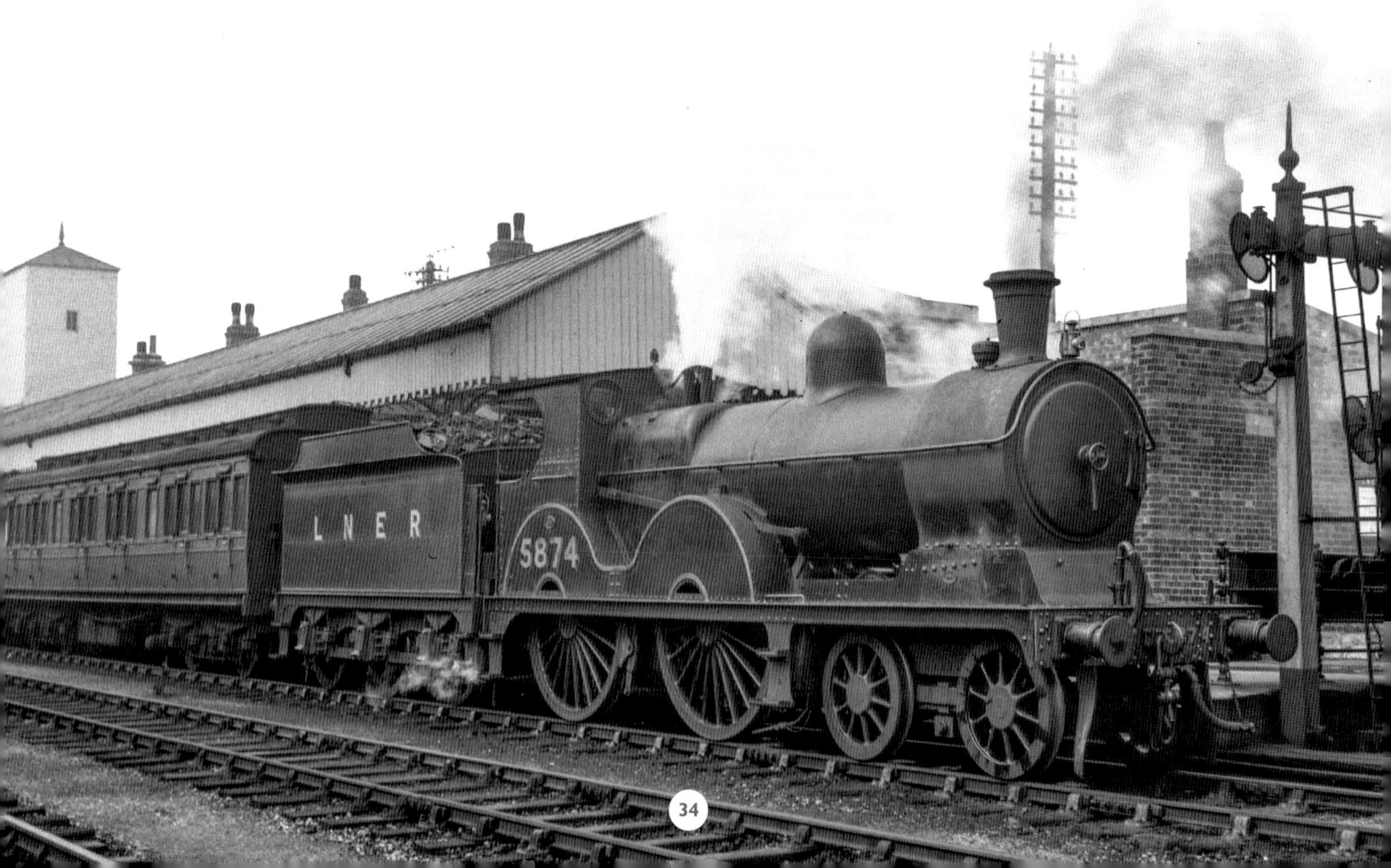
L N E R
5874

Above: J. G. Robinson designed numerous successful types of locomotive, The graceful compounds, LNER Class C4 and his Director and Improved Director 4-4-0s and, of course his 8K 2-8-0s, LNER Class O4, built in large numbers by the Railway Operating Department towards the end of World War 1. He also built a class of 0-6-0 that did sterling work on ex-GCR lines from 1901 to the end of 1962. These were the 175 locos of Class 9J, LNER J11, affectionately referred to as Pom-Poms as their staccato exhaust resembled a gun of that name.

A review of early earlier allocation records would appear to indicate that the majority of the first 40 built were allocated to Grimsby. Grimsby depot was replaced by Immingham in 1912. At the grouping, the only one of the first 30 to be built not allocated to Immingham was No. 993.

The first surviving allocation of No. 1051, by then renumbered 6051, is to Immingham in February 1926, it was to remain in the area until withdrawal in 1959. In 1933 it had a spell at New Holland before moving to Louth at the beginning of December 1939. Renumbered 4328, then 64328, it remained allocated to Louth until the depot closed in December 1956. It is approaching Habrough from the Brocklesby direction with a coal train, probably for export through Grimsby, in 1947.

Opposite above: It was common practice to find that as express engines aged and they were replaced their duties diminished and they were put onto less arduous duties at sheds distant from their original haunts. When the MS&LR Class 2 4-4-0s were introduced from 1887, they were the first single framed locos owned by that company. By the grouping, they had been supplanted on express work by successive designs and had already been put out to pasture. Many of the class found their way to northern Lincolnshire, where they would see out their days working local service form New Holland, Cleethorpes and Louth, a couple even found their way to the new shed at Frodingham, for a passenger working out of Barnetby.

Although New Holland housed 12 members of the class post-grouping, No. 5709, seen here, was not one of them. In fact, this loco had been a Cheshire Lines loco from before the grouping until 1930, coming to Immingham via Sheffield in that year. It was withdrawn from Immingham in March 1932. It was captured by the camera, during its all too brief stay in the area, waiting at New Holland Pier with a train, possibly for Immingham Dock, judging by the carriages.

Opposite below: The MS&LR introduced its Class 11A 4-4-0 express locomotives in readiness for the opening of the London extension. They were, in simple terms an improved Class 2, see above. Like the Class 2s, they were eventually replaced on express work and were a familiar sight on the Cheshire Lines for many years. A handful made it over the Pennines to northern Lincolnshire in the late 1930s, although their tenure was short.

Classified D6 by the LNER, No. 5874 was one of those, spending three years at Immingham, between May 1937 and March 1940, before returning to the Cheshire Lines. It is seen waiting departure from New Holland Town, during its time at Immingham, with an all stations train for Cleethorpes.

The J11 was a familiar sight in the area from their introduction until, virtually, the end of the class, Immingham withdrawing its final example, No. 64386 in the great cull of September 1962. As young spotters, around 1960, we thought they looked ancient with their tall chimneys, they were, indeed, old but were still capable of putting in a hard days work.

Taken the same day as that of No. 4328 on the previous page, No. 4421 is coming off the New Holland line at Habrough with, possibly, the workmen's train from Immingham Dock. This loco was never allocated to the area and was a Retford loco at the time, presumably borrowed for this duty.

The four ex-GCR Compound Atlantics came to Immingham in 1933, remaining there until withdrawal in the final year of the LNER in 1947. They were used on a variety of services from Cleethorpes and appear to have been popular with the crews. No. 5364 *Lady Faringdon* is seen in platform 3 at Cleethorpes with a stopping train for Manchester, the coaches carrying nameboards, in 1937. To the left is New Holland based Class D7 No. 5684 with a train bound for New Holland Pier. 5684 was withdrawn in 1939 whilst No. 5364, the loco that hauled the Royal Train at the opening of Immingham Dock in 1912, was the final member of the class when withdrawn on 31st December 1947.

In the background is the iconic Cleethorpes clock tower, a beacon that has guided holiday-makers back to their trains since 1880.

The loco shed at Immingham was built to service the new dock being built there and was on the south east corner of the dock estate. The 12-road structure, opened in 1912 was amongst the largest on the GCR and supplied the needs, not only of the new dock, but nearby Grimsby. The allocation was mostly centred around the goods traffic operating into and out of the docks at both ports. There was, however, a requirement for passenger engines to work the trains out of Cleethorpes.

Four Class D6's came to Immingham in 1937 to replace the ageing D7's, as mentioned above, although they only lasted two years before transfer away or withdrawal. Seen at the west end of Immingham shed in 1938 is another of the quartet, No. 5880. This loco was withdrawn in 1939. Note the large amount of coal in the tender. The water softening plant, above the coaling stage, was unique in design and can be seen on the right of the picture.

EASTERN TIMES • ISSUE 8

Above: J. G. Robinson designed nine different classes of 4-6-0 for the GCR between 1903 and 1921. Not all were successful but members of all nine classes survived until the late 1940s. Amongst the most enduring was his Class 8F, 6ft 7in driving wheel variant, introduced in 1906. There were 10 in total and they became Class B4 under the LNER. The third member of the class was the only one to carry a name. It was named *Immingham*, to commemorate the commencement of the building of the dock, the class often being referred to as *'Imminghams'*. After the grouping many of the class settled in West Yorkshire at Ardsley and Copley Hill, although one or two could still be found on GCR metals. After World War 2 the days of the GCR 4-6-0s were numbered with the introduction of Thompson's B1 4-6-0s. Nevertheless, *Immingham*, renumbered 1482 in January 1947, was turned out from Gorton in March 1947 resplendent in LNER lined green livery, the only GCR 4-6-0 to receive this treatment that late in its career.

Based at Lincoln at the time of its repaint, No. 1482 *Immingham* is seen at Barnetby on a train for its home city shortly before it transferred to Ardsley in July 1947. *Immingham* became something of a celebrity loco and was the final ex-GCR 4-6-0 in service, being withdrawn in November 1950. Another candidate for preservation? I believe so, but I'm biased towards anything ex-GCR.

Opposite above: As well as the water softening plant, another identifier to locating photographs at Immingham are the high lines in the background. These were used for taking coal wagons to the loading chutes on Immingham Dock, the wagons returned by gravity at a lower level. There was a requirement for short wheelbase shunting locomotives, especially around the docks at Grimsby. These locomotives would work at Grimsby during the week and return and were serviced at Immingham on a weekend. Pollitt designed a class of 0-6-0 saddle tanks specifically for working in the dock area at Grimsby and elsewhere where tight curves abounded. There were 12, built during 1897. J. G. Robinson added a further six to basically the same design, the notable difference being that his carried side tanks, whereas the Pollitt ones were saddle tanks. The Robinson variant, Class J63 under the LNER, spent virtually their whole lives shunting around Grimsby. The Pollitt ones, class J62, ventured much further afield and only a few were based in the area in LNER days.

No. 5890 was one of the Pollitt designed saddle tanks, it was photographed at the west end of Immingham shed in 1938. After several years away from the area, this loco returned in 1938, remaining until withdrawn in 1949.

Opposite below: Although not visible in this view, a cenotaph style coaling tower was built at Immingham in the 1930s. Adjacent to this were the storage sidings used as a holding area for locos. On 8th June 1947, three ex-GCR locos were captured at this location. On the left is the tender of Class B7 4-6-0 No. 1395 and on the right is Class C4 Atlantic No. 2915. No. 1395 was based at Gorton, being withdrawn from there in November 1948. No. 2915 was based at Leicester and would enjoy spells at Lincoln and Boston before withdrawal in June 1949. The loco in the centre of the trio, is Class B2 No. 1491 *City of Manchester*. Based at Immingham, it was withdrawn just a month after this photograph was taken. Immingham was the last home for all six B2s, withdrawal taking place between December 1944 and November 1947. The survivors were reclassified B19 in August 1945 to make way for Thompson's rebuilt Class B17s to become Class B2.

THE MAGNIFICENT CLAUD HAMILTONS

BY DAVID CULLEN

1930s • Class D16/3 4-4-0 No. 8837 and B17/1 4-6-0 No. 2836 *Harlaxton Manor* approach Chadwell Heath on the Down fast main line. The distinctive bridge seen in the background is No. 90 which is located around 10 miles from Liverpool Street station.
Photo: © David P. Williams Colour Archive/Transport Library

There will be countless more, but among them the number '1900' has the following two relevancies. It is a year, one hundred and twenty-five in the past, also the operational number of a steam locomotive constructed that year at the Great Eastern Railway's Stratford Works in east London. Generally acknowledged as having been designed by Chief Mechanical Engineer Mr. James Holden, it was in fact his Chief Draughtsman Mr. Frederick V. Russell who shouldered most of the responsibility as Mr. Holden was abroad at the time.

Eponymously named *Claud Hamilton* in honour of the GER chairman, shortly after its launch the locomotive took part in The Paris Exposition in Vincennes. There it aroused great admiration which led to the awarding of a gold medal. This was in no small part due to its livery, comprising sumptuous Great Eastern Royal Blue all over, gold-coloured linings, band adorned boiler barrel, red coupling rods and a gleaming copper chimney cap. The top halves of its four coupled wheels were enshrouded in smart valances with gold coloured outlining. Each front cover bore the locomotive's name surmounting an elaborate GER coat-of-arms. The rear valances displayed the locomotive's gold coloured number on a red background. Safe to say it was one of Britain's most handsome steam locomotives, arguably the handsomest of the day and beyond.

A fleet followed, construction of which began the same year. Forty-one were built at Stratford through to 1903, the prototype plus another forty in four orders of ten. Given the classification S46, they were quickly dubbed *Clauds*. Following the 1923 Grouping, the GER was absorbed into the London & North Eastern Railway which reclassified them as D14s. Subsequent engines built 1903–1933 were classified D15, D15/1, D15/2, D16/1, D16/2 and D16/3. Having larger boilers with increased evaporation capability, the ten D16/1s and D16/2s were known as *Super Clauds*. The sub-classes eventually totalled one hundred and twenty-one.

Beginning in 1903 many rebuilds were carried out. These were for purposes including fitting Belpaire fireboxes in D15s to replace their original round topped boxes, installing superheating on D15/1s and elongated smokeboxes on the D15/2s, replacing original slide valves with more modern 'piston' types on twenty and finally, moving full circle, reinstalling round topped fireboxes.

Technical Details

The Clauds were of the 4-4-0 wheel arrangement. This was dubbed the 'American' notation, being carried by countless locomotives opening up North America in earlier decades. With their unmistakeable spark-arresting smokestacks and tenders loaded with logs for fuel, their mere presence enhanced numerous westerns.

An undated shot of Holden Class D15 No. 8812 ready for work at an unspecified depot. Built in March 1910 the loco was initially numbered 1812 by the GER. It was withdrawn by BR in November 1948 still using its LNER 1946 number 2583.
Photo: © Transport Treasury

An undated shot of an immaculate Holden Class D15, No. 8787.
Photo: © Transport Treasury

The four Claud driving wheels were of 7ft diameter, the four bogies 3ft 9ins and the six tender wheels 4ft 1in. The tenders held 715 gallons of fuel (this statistic is covered in due course) and 3,450 gallons of water. Overall length was 53ft 4¾ins. Total wheelbase; the measurement between the leading bogie axle and the rearmost tender axle, was 43ft 8in across all sub-classes.

Weight in full working order varied between 89 tons 11 cwt for the D14s and 95 tons 3 cwt for the D16/3s. Overall maximum axle load was 18 tons 13 cwt held by the D16s. Adhesive weight for rail grip varied between the D14s' 33 tons 4 cwt and the D16/3s' 36 tons 9 cwt. The heavy fireboxes carried over the space within the driving wheels contributed to these totals and further provided stability during motion.

Boiler maximum diameters were 4ft 9ins for the D14s and 15s and a fraction over 5ft 1in for D16 developments. Evaporative surfaces varied from 1,624.38 sq.ft. for the 'Diagram 28' units carried by the D14s up to 1,731.9 sq.ft. for the 'Diagram 28A' of the D16/3s. Grate areas were mainly 21.6 sq.ft. with the D16/3s having 21 sq.ft.

Superheating was introduced on the D15/1s from 1911. Two were experimentally given 18-element 'Schmidt' type superheaters. Two received Great Western Railway 'Swindon' units, along with 'jumper-top' blastpipes to reduce exhaust steam pressure if exceeding a safe level. If excessive it lifted a top-mounted ring, reducing blast strength and preventing burning coal being drawn through the boiler tubes.

The superheaters finally decided upon were efficient and much utilised 'Robinson' types. These had 18 elements of 1.1 ins diameter with a total evaporative surface of 154.8 sq.ft. D16/3 units had almost double this with 21 x 1¼in elements having a surface area of 302½ sq.ft. Maximum boiler pressure was 180 lbs per sq. inch across all sub-classes.

Again across all, two cylinders of 19 inches diameter by 26 stroke provided power. As with so many earlier locomotives, these were set within the main frames. However, by now becoming outdated, this technique was to be superseded by outside setting of cylinders, with of course continuance for locomotives equipped with four.

All Claud valves were operated by two sets of Stephenson Link motion sharing space between the frames with cylinders, motion rods and cranked axles. This valve gear would ultimately relinquish dominance to alternatives, mainly Walschaerts which could be fitted outside, allowing easier maintenance. 'Slide' valves controlled steam flow until the D16/3s came along, these being fitted with 8 and 9½ inch diameter piston valves.

A steam locomotive's reckoned operational pressure is 85% of maximum to allow for a 15% reduction in flowing from boiler to cylinders. Working with this figure of 153 lbs per sq.in. along with cylinder and driving wheel diameters plus piston stroke, Tractive Effort comes to 17,096 lbs. Although an artificial value, this gives an accurate indication of locomotive hauling capability and was a source of great pride to railway operatives and managers.

Regarding locomotive numbering. As already covered, *Claud Hamilton* started out as No. 1900. Under the LNER it was renumbered 8900. Under another renumbering programme of 1946 it became No. 2500. GER numbers ran in reverse from 1899 for the following batch built in 1900 down to 1780 on the final constructions of 1923. The LNER renumbered these 8900 to 8780 and in 1946 they were again renumbered 2500 to 2620.

The Clauds were designed with some innovative concepts. The cabs were spacious and sheltered with two windows in each side panel and two in their fronts for forward vision, providing crews with a working environment both practical and comfortable. This followed similar consideration by Mr. Wilson Worsdell, Great Northern Railway designer of the Q1 Class four years earlier.

An ingenious recycling measure was installed for firing the D14s and the first ten D15s. Waste oil residue was burned as fuel, explaining the earlier mysterious statistic; 'the tenders held 715 gallons of fuel'. This was obtained from the GER oil-gas plant which produced gas for lighting carriages. It created a convenient method of disposal and virtually eliminated the need for coal. Virtually, because steam was initially raised using coal, as steam pressure was used for injecting the oil fuel into the firebox.

The oil-burning apparatus was patented by Mr. Holden. Pre-heated by a unit in the smokebox, oil was steam sprayed into the firebox through injectors set in the front, twelve inches above a bed of firebrick. To support combustion, these injectors additionally delivered pre-heated air through internal cones, the process producing a hot, diffused and highly flammable spray.

However, oil became more expensive and carriage lighting was converted to electricity. It was then decided it would no longer be economical to use it as locomotive fuel. Subsequently the locomotives were converted to burn coal by 1912. Some were later refitted with oil burning apparatus as a precaution against coal shortages, but this was not carried out beyond 1927.

For adjusting exhaust intensity to suit running requirements the blastpipe orifices were variable. Live and exhaust steam injectors for maintaining boiler water levels were fitted to the first seven locomotives. From the eighth onward the exhaust-operated unit was replaced by an additional live steam injector. Locomotives built earlier were correspondingly adapted.

A further innovation was the reversing gear being power operated using compressed air. As were water collection scoops for on-the-move replenishing of the tenders from troughs laid between the rails. This equipment however created a major safety issue. The locomotives were equipped with Westinghouse compressed air braking and there was the risk this could be compromised by pressure depletion through the equipment's use. A simple remedy was the fitting beneath the cab of two compressed air reservoirs, these separated by a non-return valve. The pump fed pressurised air to the first reservoir. The brakes, reversing gear and scoop were powered by the second.

Despite a relatively modest appearance, at their introduction they were the GER's largest and most powerful express engines. For many years they formed the mainstay of passenger workings, linking London with locations including Clacton, Ipswich, Norwich, Cromer and Parkeston Quay. The latter operated boat train services. In addition they proved more than capable of handling fourteen carriage trains totalling some 430 tons, these regularly comprising the 'Norfolk Coast Express'. Beginning at London's Liverpool Street station these heavy workings could run the 130 miles to North Walsham in 2 hours 39 minutes. Producing a start-to-stop average of some 49 mph, this was a quite remarkable feat.

While operated on many railways, double-heading (coupling two locomotives together for greater hauling power) was not Great Eastern policy, apart from on inter-depot movements. By the 1920s this changed and it became

Another undated shot sees Class D16/2, No. 8847 passing over an unrecognised swing bridge. If any reader recognises this and the previous two locations please write in.
Photo: © Transport Treasury

common practice on heavy morning expresses serving Clacton and Walton-on-the-Naze.

After the 1923 Grouping, D14s and D15s were mainly allocated to London and around the LNER's East Anglia territory. Their birthplace of Stratford received thirty-five. Ipswich and Cambridge received nineteen each, Norwich fifteen, Colchester six, Great Yarmouth and March in Cambridgeshire five each, Parkeston three and King's Lynn two. Lowestoft received just one. Doncaster, the farthest northern placement, also acquired just one.

The LNER was famed for its Apple Green livery so replaced the engines' GER Royal Blue accordingly. Their striking red coupling rods were superseded by rods of polished steel.

From 1924 Clauds would occasionally be seen at Kings Cross for handling special workings, none more special the Royal trains bound for Cambridge or King's Lynn. Passenger services between Kings Cross and Cambridge were expanded in 1932, when Clauds from Cambridge shed provided five expresses each weekday. Although Great Northern Railway C1 and C2 Atlantics (4-4-2 locomotives) arrived to share these workings, the Clauds held their own until 1938.

In 1939, depot reallocations gave Norwich twenty-seven Clauds, Stratford twenty-four, Cambridge twenty-three, Ipswich twelve, March ten, Colchester and Great Yarmouth eight each, King's Lynn five with Bury St. Edmunds and Peterborough East each receiving two.

Accidents occur in all areas of life and railways are no exception. The Clauds were involved in a number of incidents resulting in varying degrees of seriousness.

No. 1813 was heading a morning express from Clacton to London Liverpool Street on New Year's Day 1915. Reaching Ilford in Essex, it passed signals set at danger. Despite a signalman issuing warnings, it collided with a local passenger service. Ten persons lost their lives. Around five hundred sustained injuries of varying degrees. The official report included recommendation for an automatic train warning system to be introduced.

On 16th January 1931 No. 8781 was running light engine at Great Holland in Essex. It became involved in a head-on

Two early 1950s views of ex-LNER Class D16/3 4-4-0s. Introduced in 1938, they were rebuilds of the Hill D16/2 'Super Claud', with round topped boilers but with retained original footplating and slide valves. Below we see the footplate crew of No. 62565 having a chat while being held at a signal at King's Lynn and right No. 62601 arriving at Hunstanton. *Both photos: H. Cartwright © Transport Treasury*

collision with a newspaper train headed by LNER B12 4-6-0 No. 8578. This train had recently left Thorpe-le-Soken running against signals. Both footplate crew of the B12 and the guard died. The crew of No. 8781 both sustained injuries described as serious.

On 27th November 1934 D15/2 No. 8896 derailed at Wormley in Hertfordshire following a collision with a lorry on a level crossing. Both of the locomotive crew died. Twenty-five persons sustained shock or injuries, although fortunately most were relatively minor.

On 1st June 1939 No. 8783 was involved in a collision with a lorry loaded with straw. This occurred at the Cross Drove level crossing to the north of Hilgay station in Norfolk. Although the lorry was seriously damaged the driver was uninjured. The train derailed and collided with freight wagons in a nearby siding. Four passengers lost their lives with a further five sustaining injuries.

For all the deceased mentioned here, R.I.P.

When 4-6-0 locomotives appeared in 1913 they ousted the Clauds on main GER services. However, their relatively modest weight and axle load afforded them unrestricted access to the route reaching Wolferton, the Norfolk station serving the Royal residence of Sandringham House. Subsequently the frequent Royal train duties were allocated to Clauds, two of which were especially liveried in green and maintained in pristine condition. One of these was No. 8783, salvaged and repaired following the above incident near Hilgay.

Withdrawals began in 1945 with a single engine; No. 8866. At Nationalisation on 1st January 1948, the LNER handed over one hundred and seventeen to British Railways. Thereafter withdrawals were made each year, the largest cull of twenty-eight being in 1957. The final four departing were Nos. 62524 and 62597 along with Nos. 62604 and 62613 in 1960. None would escape the scrap-yard torches.

One Claud was in fact earmarked for restoration at Stratford Works. Unfortunately the crucial instruction was only displayed on one of its sides. Dire consequences resulted, as with bitter irony, the disposal gang approached it from the opposite side. The rest, as they say, is history.

However, a scheme is in place for creating a new locomotive, a replica of D16/2 No. 8783. This was the engine involved in the Hilgay accident and following restoration, one of the Royal duty engines. This is being planned by members of the Whitwell & Reepham Railway in Norfolk. The venture is expected to take a minimum of ten years. Metaphorically rising from the ashes, the locomotive is to be named *Phoenix*. As a lifelong steam fanatic, I wish the project every success.

Information Sources:
The Great Book of Trains by Brian Hollingsworth and Arthur Cook
www.gersociety.org.uk   www.lner.info
www.newprincegeorgesteam.org.uk
www.preservedbritishsteamlocomotives.com
www.railwaysarchive.co.uk
www.railwaywondersoftheworld.com   www.skrimarket.com
www.steamlocomotive.com   www.tmb.fenhistory.uk
www.en.wikipedia.org   www.youtube.com   www.google.com

D16/3 No. 62588 ready to depart from King's Lynn and about to pass classmate No. 62518. *Photos: Dr. Ian C. Allen © Transport Treasury*

6th June 1952 • Class D16/3 No. 62576 at Saxmundham.
Photo: Neville Stead Collection © Transport Treasury

3rd October 1959 • Class D16 No. 62570 passes the fine array of chimneys at Kings Dyke brickworks near Whittlesey. The final three chimneys were demolished by controlled explosion in 2021. *Photo: Neville Stead Collection © Transport Treasury*

3rd April 1956 • Class D16/3 No. 62553 makes a spirited departure from Ely.
Photo: Peter Hay © Transport Treasury.

An undated shot of Class D16/3 No. 62525 at Wisbech station.
Photo: Neville Stead Collection © Transport Treasury

EASTERN TIMES • ISSUE 8

12th July 1959
D16/3 Class 4-4-0 No. 62613 at Fenchurch Street station with The Locomotive Club of Great Britain *The Eastern Counties Limited* Rail Tour.

The route of the outward journey was as follows:
London Fenchurch Street • Gas Factory Junction
Bow Junction • Carpenters Road Junction
Channelsea Junction • High Meads Junction
Loughton Branch Junction • Lea Bridge
Copper Mill Junction • Tottenham South Junction
Seven Sisters Junction • Bury Street Junction
Cheshunt • Broxbourne • Bishops Stortford
Audley End • Shepreth Branch Junction
Cambridge

Photo: Charles Firminger
Online Transport Archive/The Transport Library

HATFIELD LOG
THURSDAY 30TH MARCH 1961

BY GEOFF COURTNEY

For the past five issues of Eastern Times I have written about detailed logs I made between April 1957 and August 1961 of trains passing through my home station of Ilford, on the Liverpool Street–East Anglia main line. My trainspotting journey through that period, however, took me to a number of other Eastern Region locations, and the notes I made will feature in future issues, starting in this issue with a record of every train that entered my log book at the ECML station of Hatfield over a period of nearly six hours on 30th March 1961.

Without wishing to disrespect the Ilford scene, it has to be said that the diet on offer on the ECML was a sumptuous banquet that sated the appetite of any hungry trainspotter, wherever their loyalty lay. It was of Premier League quality, and in those six hours I was at Hatfield I recorded five classes of Pacifics built from 1922 through to the BR era, and eight titled trains. There was also enough of a hint that the immediate motive power future lay with diesel traction to concentrate the teenage minds of we spotters to 'enjoy it while you can,' but more of that anon.

What strikes me 64 years after that late-winter's day is the importance to successful operations of the oldest express locomotives on display, Nigel Gresley's A3 class, of which no fewer than 13 went into my notebook, nearly as many as the combined total of the A1, A2 and A4 representatives. The oldest of these veterans was No. 60102 *Sir Frederick Banbury*, which had been outshopped by Doncaster nearly four decades earlier, in July 1922, and to add salt into the wounds of the far younger motive power that was strutting its stuff that day, this engine had been tasked with heading the prestigious 'Heart of Midlothian' King's Cross-Edinburgh Waverley working.

It is also worth mentioning at this juncture that No. 60102 was one of no fewer than five A3s logged in the space of just 18 minutes, the other four being Nos. 60111 *Enterprise*, 60105 *Victor Wild*, and 60109 *Hermit*, on Up expresses, and No. 60054 *Prince of Wales* en route to Leeds.

Still on the A3 theme, No. 60112 *St. Simon* sneaked into my logbook twice, once running light with V2 class 2-6-2 No. 60896 on the Up line (can two engines coupled together be described as running light?), and passing me nearly 3½ hours later in charge of a Down express. Indeed, my notes show that the V2 also reappeared on a Down express, leading me to wonder if this was a regular operational manoeuvre, or whether the pair, allocated respectively to Grantham (34F) and Doncaster, (36A), had travelled south in answer to a motive power SOS from Top Shed.

Another member of the class, No. 60049 *Galtee More*, was entrusted with the Down 'Northumbrian' to Newcastle, and an 11-minute early afternoon spell resulted in Nos. 60056 *Centenary*, 60046 *Diamond Jubilee*, and 60039 *Sandwich*, entering my log on a trio of expresses. Before moving on to other worthy locomotives, I will end this A3-fest with an A3 that wasn't an A3, No. 60113 *Great Northern*, which early in my visit came past with the Up 'West Riding' from Bradford and later in the day on a Down train to Leeds.

This Pacific started life in April 1922 as the prototype of the Gresley-designed A3 class (which just to confuse matters was at that time the A1 class), but in 1945 was chosen by his successor, Edward Thompson, to be rebuilt as the prototype of his Class A1. The decision by Thompson to use what was Gresley's first Pacific locomotive for this role has been the subject of much debate over the past 80 years, but that is for another time.

Talk of Gresley inevitably leads to thoughts of his iconic A4 class, but on this day, they were very few and very far between. In fact, just four passed me by, comprising No. 60034 *Lord Faringdon* working the Down 'Tees-Thames Express,' No. 60006 *Sir Ralph Wedgwood* on an Up Newcastle train, No. 60017 *Silver Fox* on the afternoon Scotch Goods fast freight service from King's Cross to the Edinburgh suburb of Niddrie, and would you believe, No. 60033 *Seagull* on an Up parcels train. Another Gresley-designed named engine on view was the now preserved V2 class pioneer No. 60800 *Green Arrow*, named after an LNER express parcels service but on this day on a Down Edinburgh train.

Right: **15th August 1959 • A1 No. 60148** *Aboyeur* is at Grantham on the 'Yorkshire Pullman'. Geoff Courtney logged this Pacific at Hatfield on the same working on 30th March 1961.
Photo: Arthur W. Cundall © Transport Treasury

19th March 1955 • The pastoral setting of Hertfordshire is disturbed as A1 No. 60117 *Bois Roussel* bursts out of Welwyn North Tunnel with 'The Queen of Scots' Pullman express, a combination identical to that recorded by Geoff Courtney at Hatfield almost exactly six years later. *Photo: Roy Edgar Vincent © Transport Treasury*

This class of 2-6-2 locos was designed for mixed traffic work, and this was clearly evident on my visit, as the 12 that I logged were on everything from express operations to semi-fasts, and from freight to parcels and empty stock.

Another class that was determined to make its presence felt was the post-Edward Thompson A1 designed by Arthur Peppercorn in the last days of the LNER and built by BR at Doncaster and Darlington in 1948-49. Ten of them entered my notebook, among which were No. 60117 *Bois Roussel* on the Down 'Queen of Scots' Pullman to Glasgow via Leeds, No. 60158 *Aberdonian* 12 minutes later on the Up 'Tees-Thames Express,' and No. 60148 *Aboyeur* in charge of the Up 'Yorkshire Pullman.'

Indeed, traffic was, as usual, frequent between the capital and Yorkshire, and my log shows that A1s based in that county operating Up or Down services to or from York or Leeds also included Doncaster (36A) residents Nos. 60139 *Sea Eagle*, 60149 *Amadis*, and 60157 *Great Eastern*, and No. 60115 *Meg Merrilies* of Leeds Copley Hill (56C).

Within the first minute of my arrival at Hatfield I was able to log three locomotives, the first being Class 9F 2-10-0 No. 92186 of Peterborough New England (34E) on freight duty, immediately followed by two other New England residents, A2 No. 60500 *Edward Thompson* and V2 No. 60914 on their way light engine to Kings Cross, both to return later in the day on Down expresses.

That early viewing of the Pacific made me think other members of the A2 brotherhood would go into my logbook, but that was not to be. Doncaster May 1946-built No 60500 was the first of the day, and the last, despite a clutch of them being allocated at that time to either New England or Doncaster.

One other Pacific I must give a shout-out to is 'Brit' No. 70040 *Clive of India*, which only three months before this sighting had been transferred from operations on my local Liverpool Street to Norwich main line and which, now an Immingham (40B) locomotive, was heading an express from Cleethorpes.

17th June 1961 • No. 70040 *Clive of India* hustles a Cleethorpes-Kings Cross working through Hatfield, just weeks after the Standard Pacific has been logged on a similar train at the same location. *Photo: Leslie Freeman © Transport Treasury*

Despite the diesel revolution that at this stage was rapidly gaining momentum, steam still held sway on freight trains, and led diesels 9-6 with a staple diet of V2, 9F and WD 2-8-0 motive power.

My visit to Hatfield came a few months before the Deltics (later Class 55) brought their 3300bhp to the ECML revenue-earning service, so the flagship diesels I logged were in the D200 Type 4 (later Class 40) series. Top of the charts in this regard were D208 on the Down 'Flying Scotsman' and D207 on the Up 'Tees-Tyne Pullman,' and among four other members of the class that went into my notebook was D279 that made an appearance on a Down freight train, ironically passing by two minutes after another elite locomotive, No. 60033 *Seagull*, on the aforementioned parcels train.

One of the enigmas of that time, to me at least, was the 'Cambridge Buffet Express.' As a King's Cross and ECML regular I was aware of its existence, but although I have given it a title in my logbooks, I don't recall ever seeing a headboard, and nor is it listed as a named train in the contemporary Bradshaw's Guides that still grace my bookshelves.

Perhaps I identified it by carriage boards, although I don't recall them, and also, I wonder if, as it was a regular working throughout the day, it was multi-portioned and thus the 'Atlantic Coast Express' of the Eastern Region. Whatever its background, I logged this train six times on the day, equally shared between Up and Down, and motive power on five of them was Type 2 D55xx (later Class 31) diesels, the sole exception being another Type 2, 'Baby Deltic' D5907 (later Class 23).

And so this day of ECML superpower came to an end. Further ECML logs from the steam/diesel transition era will feature in future issue of Eastern Times, recorded at such locations as Peterborough North, Wood Green, Hitchin, Hadley Wood, and Stevenage.

6th May 1961 • Gresley power in the shape of A3 No. 60049 *Galtee More* displays its grace as it runs though Hatfield with the Down 'Northumbrian', six weeks after Geoff Courtney had noted the same locomotive on the same express also at Hatfield.
Photo: Peter Pescod © Transport Treasury

EASTERN TIMES • ISSUE 8

Above: **17th September 1960** • There's nearly 40,000lb-ft of tractive effort available to this unfitted goods train as No. 92186 passes Hatfield. The Class 9F 2-10-0, one of the most powerful locomotives in the BR fleet at the time, was also logged at Hatfield on 30th March 1961, again on freight operations. *Photo: Ron Smith © Transport Treasury*

Below: Controversial Class A1 prototype No. 60113 *Great Northern* reverses out of Kings Cross on an unknown date, probably in the late-1950s, having brought in the *West Riding* express. Another member of the class, No. 60139 *Sea Eagle*, which inherited its name from A4 No. 4487/BR No. 60028, simmers in the left background, and just visible behind No. 60113 is what appears to be a D55xx (later Class 31) diesel. No. 60113 was a familiar sight on the 'West Riding', including at Hatfield on 30th March 1961. *Photo: © Transport Treasury*

13th September 1958 • Class pioneer 2-6-2 No. 60800 *Green Arrow* passes Wood Green Up No. 4 signal box with a Kings Cross-bound express while N2 class 0-6-2T No. 69531 awaits the right away at a suburban platform with a Finsbury Park train. The now preserved V2, named after an LNER express parcels service, was recorded by Eastern Times contributor Geoff Courtney on a Down Edinburgh express at Hatfield on 30th March 1961. *Photo: R. C. Riley © Transport Treasury*

1949 • A4 Pacific No. 60017 *Silver Fox* is in fine fettle at Potters Bar with the 'Yorkshire Pullman.' More than a decade later, on 30th March 1961, the Pacific was logged five miles up the line at Hatfield in charge of the afternoon Scotch Goods fast freight service to Edinburgh.
Photo: Neville Stead Collection © Transport Treasury

Above: **10th June 1950** • It's straight ahead for No. 60500 *Edward Thompson* with a Down express at Abbots Ripton, between Huntingdon and Peterborough. This locomotive was one of a number of Pacifics logged by Geoff Courtney during a six-hour trainspotting visit to Hatfield on 30th March 1961. *Photo: Roy Edgar Vincent © Transport Treasury*

Below: **19th September 1958** • Type 4 D208 is ready for action at Kings Cross with the 5.35pm to Newcastle, while N2 Class 0-6-2T No. 69594 represents the old guard in the background on empty stock duty. The Type 4 diesel was logged at Hatfield on 30th March 1961 in charge of the Down working of the ECML's flagship express, *The Flying Scotsman*. *Photo: Alec Swain © Transport Treasury*

29th July 1952 • A double delight for photographer Eric Sawford as A2 Pacific No. 60500 *Edward Thompson* and V2 No. 60800 *Green Arrow* meet up at Huntingdon, respectively on ECML express and freight operations. Each of this pair entered Geoff Courtney's log of trains at Hatfield on 30th March 1961 on Down expresses.

Photo: Eric Sawford © Transport Treasury

THE THOMPSON CLASS L1 2-6-4 TANK LOCOMOTIVES

21st June 1957 • Class L1 No. 67765 pictured at Whitby West Cliff.
Photo: Neville Stead Collection © Transport Treasury

Edward Thompson (1881-1954) was Chief Mechanical Engineer of the London and North Eastern Railway between 1941 and 1946.

He came from an academic background having taken Mechanical Science Tripos at Pembroke College, Cambridge. This is unlike his predecessor Nigel Gresley who gained practical experience as a pupil at Horwich Works. After graduation Thompson was working in both industry and the railways for a while, when in 1912 he was appointed Carriage and Wagon Superintendent for the Great Northern Railway, a role he remained in for 18 years. In 1930 he became Workshop Manager at Stratford Works before becoming CME of the LNER in 1941 after the untimely death of Gresley. It is widely known that Gresley and Thompson disagreed on a number of issues. Many have interpreted this bitterness as NER (Thompson) vs. GNR (Gresley). Thompson also happened to be the son-in-law of former NER CME, Sir Vincent Raven. The biggest disagreement between the two was on Gresley's three-cylinder conjugated valve gear. While this valve gear arrangement worked well during peacetime, it experienced problems due to low maintenance during World War 2. This did give Thompson some justification for the criticism of the design.

Thompson instigated a much needed standardisation programme. This programme demonstrated Thompson's dislike for Gresley's engineering practices. The most controversial was the A1 chosen for rebuilding which was none other than *Great Northern*, this being the original Gresley prototype for the class. It has been stated that this rebuild was sheer vindictiveness on Thompson's part towards his former boss.

After taking up the role of CME in 1941, Thompson quickly produced a standardisation programme that listed the existing V3 2-6-2T as the standard heavy passenger tank engine, replacing the A5 4-6-2Ts, N2 0-6-2Ts, and some N5 0-6-2Ts. Later, the V3 requirement was replaced with a new two-cylinder 2-6-2T with smaller wheels and a tractive effort of 32,080lb, it was suggested that this new locomotive might have to be a 2-6-4T. By January 1943, the Running Department had requested a modern version of the London Passenger Transport Board (LPTB) L2 2-6-4T. The order for the V3s were cancelled or postponed, and plans were prepared for the new 2-6-4T. The boiler was based on the V3 but with a larger firebox and longer water tanks. Attempts were made to widen the boiler diameter but this proved impossible without reducing the water capacity. A 5 foot parallel boiler was used after a tapered design was prepared, but rejected by Thompson.

Thompson departed from Gresley's preference of a double swing link suspension on the pony trucks. He chose a design using helical control springs copied from the LMS-designed O6 (8F) 2-8-0. A similar spring design was fitted to the V2 2-6-2s after poor maintenance of their double

29th June 1947 • No. 9000 at Neasden shed.
Photo: Neville Stead Collection © Transport Treasury

22nd May 1948 • Thompson Class L1 No. 9002 at Stratford. *Photo: Neville Stead Collection © Transport Treasury*

swing link suspension mechanisms had caused a series of derailments. The bogie design was similar to that of A1/1 *Great Northern* which was also being rebuilt at the same time. The cylinders were of the same design as the B1.

Thirty L1 locomotives were authorised in April 1944, the first of the class – No. 9000 – was completed in May 1945. The LNER's 1945 Modernisation Plan included a total of 110 locomotives, including No. 9000, which would remain the only L1 in service until Nationalisation in 1948. The remaining 29 locomotives from the original order were built at Darlington Works. The plan for 110 was subsequently reduced to 100, with the remaining 70 built by the North British Locomotive Co. and Robert Stephenson & Hawthorn, between 1948 and 1950.

No. 9000 was completed in May 1945, and immediately started a comprehensive series of trials that continued into early 1947. These trials are thought to be the most extensive ever arranged by the LNER for a tank engine, and included hauling express passenger services to heavy shunting. Initial trials were based at Stratford and included suburban services out of Liverpool Street. It easily performed these duties and recorded lower coal consumption figures than the various comparison locomotives. The GE Section also reported that the L1 rode better than any other class that they were using, and the view from the cab when running backwards was an improvement over that of the V3. The GN and GC Sections also ran passenger trials with No. 9000 with similar results. No. 9000 was trialled on freight workings usually performed by J50 0-6-0Ts, K3 2-6-0s, O4 2-8-0s, and Q4 0-8-0s. The GN Section O4 and Q4 duties were handled satisfactorily, but No. 9000 had severe problems holding heavy trains on gradients on the GC Section. This is not surprising, as a heavy goods locomotive with tender such as an O4 typically had about double the braking capacity of an L1 tank engine.

The production locomotives received a mixed response, there were complaints about draughty cabs and modifications were applied to some of the L1s in an attempt to improve this. This led to all of the remaining L1s receiving similar modification between 1955 and 1960. The side tanks were welded and tended to leak. No. 9000 had tank repairs in December 1945, and the production locomotives also required a number of tank repairs. Extra stays were added to some, and from 1957 these were recommended for all remaining locomotives. The production locomotives were also fitted with water tank fillers close to the cab. This was because it was impossible to fill No. 9000 from some platform-based water columns, if the locomotive had backed into the station bunker-first.

May 1945 • Another view of No. 9000 on its launch day at Liverpool Street station. It was the only tank engine built by the LNER that was painted in lined apple-green livery.

I was sent this quote from Eastern Times contributor Dave Brennand who is an admirer of the Class L1: "I know quite a lot of drivers who worked on the L1's at Stratford and despite trials saying that they rode well, amongst the men who worked on them they were known as *Cement Mixers* because of rough riding."
Photo: © *Transport Treasury*

No. 67714 at Chalfont & Latimer station on a local working.
Photo: Neville Stead Collection © Transport Treasury

September 1951 • No. 67795 pictured at Norwich Thorpe.
Photo: Jim Flint/Jim Harbart © Transport Treasury

The axleboxes on the L1 also tended to overheat. A reason for this is thought to be the sharp curves at the end of the platforms at terminal stations which placed undue stress on the frames. Water from the previously mentioned leaking side tanks is also thought to have contributed to axlebox wear. To reduce this, five of the Neasden L1s had liners fitted to their cylinders in May 1951. These liners reduced the cylinder diameters from 20 inches to 18¾ inches and the tractive effort from 32,080lb to 28,180lb. At first this appeared to be successful, but the other four L1s continued to suffer from axlebox problems and the experiment was not deemed to be successful, however the liners were not removed. In March 1953, Neasden converted five L1s to run at 200psi rather than 225psi, another attempt to reduce the piston load and hence the axlebox wear. This experiment also failed, and by 1955 some of the L1s were described as being in a 'deplorable' condition with a whole host of problems associated with axlebox wear. This second experiment finished in October 1955 and all five locomotives were reverted back to 225psi.

After the trials, No. 9000 ran for a few months in the Edinburgh area before entering service at Stratford. The first batch of 29 production L1s were allocated to Stratford, Neasden, and Botanic Gardens (Hull). Six of the next 70 were allocated to Eastfield for use in Scotland. However, their routes were restricted by electrification in the Glasgow area and they were sent to Eastfield in exchange for V1 and V3 2-6-2Ts. After all one hundred L1s were delivered at the end of 1950, they were concentrated at Stratford, Neasden, and Ipswich; with smaller numbers at Kings Cross, Hornsey, Hitchin, Norwich, Lowestoft, Darlington, and Botanic Gardens. This would be the basic distribution until the first diesel multiple units were introduced in 1955. The L1s generally hauled passenger services in the areas operated by these sheds. Suburban and stopping passenger services were the most common, but occasionally they would haul express services, as well as some freight and colliery traffic. Over time they would replace older tank locomotives still in operation, and were in turn displaced by diesel multiple units and electrification.

Withdrawals started in 1960 and were quick. The last L1 was withdrawn in 1962. Their withdrawal was forced by the rapid introduction of diesel multiple units, rather than perceived design deficiencies.

Specification	
Cylinders (outside) x2	20in. x 26in.
Walschaerts Motion	10in. Piston Valves
Boiler: Diameter	5 feet
Boiler: Pressure	225psi
Boiler: Diagram Number	115
Heating Surface: Total	1,620.5 sq.ft.
Heating Surface: Firebox	138.5 sq.ft.
Heating Surface: Superheater	284 sq.ft. (22in. x 1.244in.)
Heating Surface: Tubes	830 sq.ft. (150in. x 1.75in.)
Grate Area	24.74 sq.ft.
Wheels: Leading	3ft. 2in.
Wheels: Coupled	5ft. 2in.
Wheels: Trailing	3ft. 2in.
Tractive Effort @ 85% boiler pressure	32,080lb
Total Wheelbase	34ft. 6in.
Engine Weight	89 tons 9 cwt
Maximum Axle Load	20 tons
Coal Capacity	4 tons 10 cwt
Water Capacity	2,630 gallons

25th March 1961 • Class L1 No. 67765 pictured at Eaglescliffe with a parcels working. *Photo: Neville Stead Collection © Transport Treasury*

Class L1 No. 67764 on a local service at Ravenscar, North Yorkshire.
Photo: Neville Stead Collection © Transport Treasury

21st June 1957 • Class L1 No. 67737 at Brimsdown in the London Borough of Enfield with a suburban service. *Photo: Neville Stead Collection © Transport Treasury*

Above: **1961** • No. 67710 ready to depart Manchester Piccadilly with a parcels working.
Photo: Jim Flint/Jim Harbart © Transport Treasury

Below: **23rd July 1958** • The final member of the L1 Class, No. 67800, pictured on a freight working comprising wooden planked open wagons at Grantham. *Photo: Neville Stead Collection © Transport Treasury*

L1 ON STAITHES VIADUCT

The viaduct featured below straddled Staithes Beck in Yorkshire, being situated north of Staithes railway station. Major crossing structures, including the viaduct, on the Whitby to Loftus line were constructed of iron, with the piers additionally filled with concrete. Designed by John Dixon, the Whitby, Redcar and Middlesbrough Union Railway (WR&MUR) commenced the building of the viaduct in 1875, although it did not open until 1883 – this was due to financial, build and ownership problems.

The viaduct was constructed from tubular iron filled with concrete, with seventeen spans; six spans of 60 feet in the middle of the bridge, and a further combined eleven spans at either end of the bridge measuring 30 feet each. The bridge was 790 feet long, and was elevated 152 feet above Staithes Beck, with one of the piers sunk into the riverbed. The piers of the viaduct were constructed of tubular steel, filled with concrete. As built, the viaduct did not have the strengthening spars running horizontally through the piers; these were added eight years after opening, with some stating that it was a reaction to the Tay Bridge disaster (see pages 21–27).

The coast routes from Whitby were deemed to be awkward to build in terms of geology, necessitating large engineering programmes. The WR&MUR line was abandoned by the original contractors due to financial problems, and the NER took over the line, but had to effectively rebuild many of the tunnels and bridges. The viaduct at Staithes was no exception; the tubes of the piers were supposed to have been filled with concrete, and when they were opened up, it was found that only gravel had been poured into the tubes. So concrete made from local sandstone mixed with Portland cement was inserted as per the original intention. These extra works further delayed opening of the line by two years, with the line opening in 1883 instead of 1881. The line was assessed at least twice by a government inspector, with various recommendations for improvement of works. One report submitted by Major-General Hutchinson noted defects in at least three of the piers of Staithes Viaduct, and was also the first person to mention a wind gauge and possible speed restrictions.

The ironwork for the viaduct was constructed off-site at the Skerne Iron Works, in Albert Hill, Darlington. The same company provided all the ironwork for the other four viaducts on the Whitby-Loftus line. The viaduct at Staithes was the tallest and longest, often being described as "spectacular". The cross-sections of iron were also fabricated by the Skerne Works and were of heavy iron bars. The diameter of the pier tubes on the 30 foot spans was 2 feet 6 inches, with the same thickness at the top of the 60 foot spans, however they tapered to 4 feet 6 inches at the bottom.

To measure wind speed and direction the North Eastern Railway installed an anemometer on the viaduct in 1884 that was designed to ring a bell in Staithes signal box should the force of the wind reach a pressure greater than 28 pounds per square foot (1.3 kPa). This would then prompt a track investigation. In March of the same year instructions were issued that the line speed across the viaduct was 20 mph and that if the storm rang the bell in the signal box, all effort should be made to stop southbound trains travelling over the viaduct. However, northbound trains were allowed to draw into, and wait, at the station. In 1935, the LNER stated that the system had hardly been used, however, because of concerns about the winds across the viaduct, the equipment was replaced, but not until 1946. The last windspeed anenometer used on the viaduct is now in the collection of the National Railway Museum in York. The line closed in 1958 and the viaduct demolished in 1960.

Below: A panoramic view of Class L1 2-6-4T No. 67754 crossing the viaduct. Photo: Neville Stead Collection © Transport Treasury

STOCKTON SHED

As we celebrate 200 years of the modern railway in 2025 we thought it appropriate to have a look at a location involved right at the beginning of the modern railway – Stockton. We've opted to illustrate the less glamorous, but essential, environs of the locomotive shed.

In July 1890 authority was given for a new station and engine shed at Stockton, the estimated cost being £48,500. The site selected for the shed was that formerly occupied by the West Hartlepool Railway workshops at North Shore Junction located to the north of the station. In 1891 the North Eastern Railway had completed the eight-road brick built dead-end shed and coal stage at a cost of £13,000. Installation of a 50ft turntable had also been authorised, this was to be situated to the east of the shed. Although a turntable was installed, it measured 46ft 6in not 50ft.

As a result of the depression of the 1930s Stockton shed and the goods yards suffered loss of work and there were closures of some freight facilities, this meant an allocation of just 19 engines were required. However, war in 1939 saw an expansion and rejuvenation of Stockton goods yards, as the railways had more traffic than they could handle. By August 1950 the shed's allocation had risen to more than fifty engines ranging from small shunting engines to heavy freight and mixed traffic locos. As well as local goods and passenger workings the shed's engines also travelled further afield to York, Whitby, Newcastle and the Leeds area. Often larger express engines would visit Stockton, but as the turntable at the shed could not accommodate them it was necessary to send them to Norton Junction to turn on the triangle.

Stockton shed closed on 14th June 1959, although there were still 31 steam locomotives allocated there, they subsequently moved to Thornaby Depot who took over the shed's work. The closing of the goods yards occurred in early 1960s with the opening of Tees Yard.

8th August 1948 • Wilson Worsdell ex-NER Class D20/1 No. 2390 at Stockton Shed. *Photo: Neville Stead Collection © Transport Treasury*

8th July 1956 • Thompson Class B1 4-6-0 No. 61019 *Nilghai* stabled at Stockton shed with a pair of vintage engineers coaches. *Photo: Eric Sawford © Transport Treasury*

STOCKTON SHED TRACK PLAN c. 1928

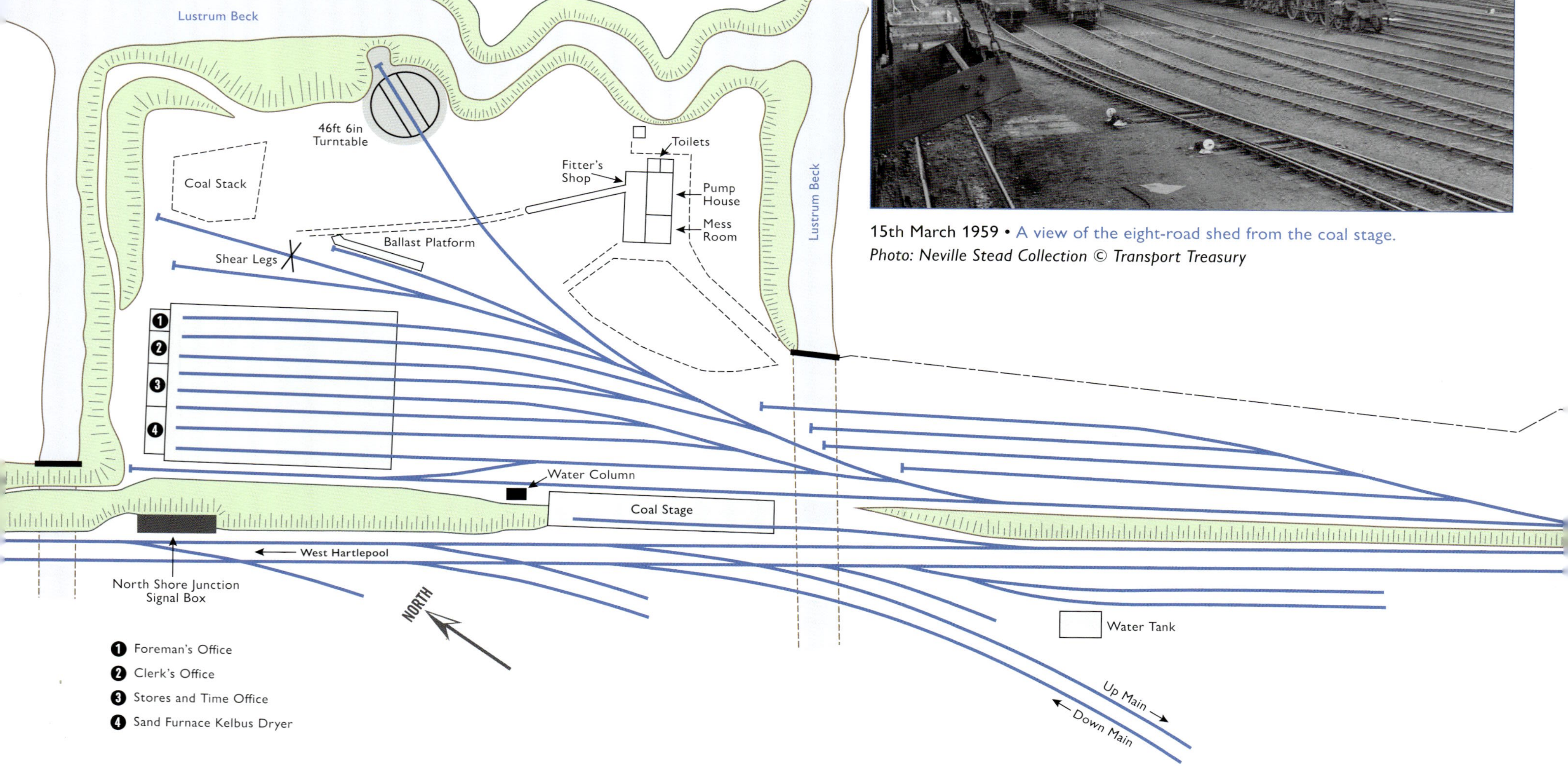

15th March 1959 • A view of the eight-road shed from the coal stage.
Photo: Neville Stead Collection © Transport Treasury

- Lustrum Beck
- 46ft 6in Turntable
- Coal Stack
- Toilets
- Pump House
- Mess Room
- Fitter's Shop
- Ballast Platform
- Shear Legs
- Water Column
- Coal Stage
- Water Tank
- West Hartlepool
- North Shore Junction Signal Box
- Up Main
- Down Main
- NORTH

① Foreman's Office
② Clerk's Office
③ Stores and Time Office
④ Sand Furnace Kelbus Dryer

8th July 1956 • Peppercorn Class K1 2-6-0 No. 62064 with classmate No. 62043 behind on Stockton shed. *Photo: Eric Sawford © Transport Treasury*

15th March 1959 • Stockton Depot, coded 51E in the Darlington District of BR (NE Region), contributed power mainly for the intensive freight traffic of the Teesside area. In 1954, Stockton had an allocation of 53 locomotives: Eleven 4-6-0s, fifteen 2-8-0s (all ex-WD), three 2-6-0s, five 0-8-0s, five 0-6-0s, one 4-8-0T, two 4-6-2Ts, four 0-6-0Ts, six 0-4-4Ts and one 0-4-0T.

The main line to West Hartlepool and Sunderland is on the left, the signalbox being North Shore. Note the two young lads kicking around as the photographer manages to get this wonderful panoramic shot of the shed on a quiet Sunday afternoon. The only two locos recorded are Class K1 2-6-0 No. 62001 on the left behind the post and Class B1 4-6-0 No. 61220 to the right of the clock.

Photo: Neville Stead Collection © Transport Treasury

THE HEADSHUNT

In future issues our aim is to bring you many differing articles about the L.N.E.R., its constituent companies and the Eastern and North Eastern regions of British Railways. We hope to have gone some way to achieving this in previous issues.

Eastern Times welcomes constructive comment from readers either by way of additional information on subjects already published or suggestions for new topics that you would like to see addressed. The size and diversity of the L.N.E.R., due to it being comprised of many different companies, each with their differing ways of operating, shows the complexity of the subject and we will endeavour to be as accurate as possible but would appreciate any comments to the contrary.

We want to use this final page – The Headshunt – as your platform for comment and discussion so please feel free to send your comments to: tteasterntimes@gmail.com or write to Eastern Times, Transport Treasury Publishing Ltd., 16 Highworth Close, High Wycombe HP13 7PJ.

GRESLEY N2

Good morning Peter

I was interested to read David Cullen's article in ET No 6 about the N2s.

Although a small point, for the sake of historical accuracy, it should perhaps be noted that Nigel Gresley was Locomotive Engineer of the Great Northern Railway, not Chief Mechanical Engineer when the N2s were introduced. There has been some speculation over the years as to why the GNR did not adopt the title, which was widely used by other railways, but it was only introduced with the coming of the LNER. As is well known, there were a number of possible contenders from the various pre-grouping companies for the post of LNER CME to which Gresley was ultimately appointed.

Mention is made in the article to the Gresley Society Trust's N2 No 1744 (69523). The loco, which is the oldest in existence to the design of Sir Nigel Gresley, is coming towards the end of its most comprehensive (and expensive) overhaul in the preservation era. Hopefully, she will be running again on the North Norfolk Railway later in 2025. The Trust are currently fund raising to help pay for the overhaul and more details can be found on our website (www.gresley.org).

Kind regards
Philip Benham MBE FCILT
Chairman, The Gresley Society Trust

MORE GRIMSBY TRAMS

I have found two postcards that may be of interest to Paul King. I am enjoying Eastern Times, particularly the features on lesser known lines.

Thanks and kind regards,
David Rockliff, Knaresborough, North Yorkshire

ISSUE NO. 7

Dear Sir

I've just received and read the above Issue, and fascinating it is too, especially the article on the Grimsby trams. I have already read articles about these trams over the years, but one thing has always stuck in my mind. The trams seemed to be very long and slim. Was the narrow body designed to enable the trams to pass one another on tight bends?

They were very attractive vehicles and it is sad that so few survived into preservation. By the way, at least one of the Cruden Bay trams survives, beautifully preserved, in the Alford Transport Museum.

An excellent magazine.

Regards, Ken McKee

UNFORGIVEABLE!

Dear Sir,

I have just received the latest copy (issue 7) of Eastern Times, which hitherto I have regarded as an excellent and well-informed publication. Regrettably my illusion has now been shattered by your apparent inability to distinguish between N2 and V3 locomotives (see back cover). Does nobody proofread the magazine?

Peter Spence (very disappointed customer).

Hello Peter

Thank you for your observation, all I can do is apologise unreservedly for the error. The books are proofread but mistakes happen from time to time, admittedly this was a bad one.

I will publish your letter in ET8 by way of apologising.

Best wishes
Pete Sikes, Editor, Eastern Times

Hi Pete,

Just to say thanks for your prompt response. It is good to know you care!

Kind regards,
Peter Spence